Time is Gravity

Volume One

TIME, GRAVITY, AND THE ORIGINS OF REALITY

GERALD CLERGE

MILTON & HUGO L.L.C.
4407 Park Ave., Suite 5
Union City, NJ 07087, USA

Website: *www. miltonandhugo.com*
Hotline: *1- 888-778-0033*
Email: *info@miltonandhugo.com*

Ordering Information:
Quantity sales. Special discounts are granted to corporations, associations, and other organizations. For more information on these discounts, please reach out to the publisher using the contact information provided above.

Library of Congress Control Number: 2025921356
ISBN-13:   979-8-89285-698-0   [Paperback Edition]
                  979-8-89285-699-7   [Digital Edition]

Rev. date: 11/17/2025

"The Foundations of Belief and Perception"

Chapter 1.1: The Genesis of Belief
The narrative commenced with an individual who witnessed a deity that profoundly influenced humanity. The consequences of Moses' revelation were extraordinary. It began when an ancient figure made an exceptional assertion of having encountered God and provided a remarkable description.

Chapter 1.2: The Turning Point
Morpheus' speech was crucial for both Neo and the movie. The blue pill means returning to an ordinary life as Thomas Anderson, but the red pill reveals the truth about the world. Choosing the blue pill maintains normalcy. Continue reading and take the red pill; this book can transform your life.

Chapter 1.3: Moments of Impact
I have a theory about moments. My approach considers the significant moments of impact—high-intensity events that can drastically change our lives. These moments define who we are and can affect us in ways we cannot predict, leading to unexpected ventures. The nature of these moments is such that it is challenging to control how they will influence us. It is often best to allow the events to unfold naturally and observe the outcomes.

Chapter 1.4: The Sense of Life
Some human beings possess a different 'sense of life' than you or me; therefore, they have a different explicit philosophy. A person's sense of life is his subconscious, emotionally integrated view of existence; it

represents his basic, early value integrations, which determine his adult - (i.e., conscious, explicit) philosophy of life. Very few of these individuals learned to value truth, knowledge, science, ethics, and justice early on. As a result, they've wrecked the functioning of their cognitive mechanism; they can't correctly identify or evaluate the facts of reality, i.e., they can't think. These individuals are primarily motivated by the irrational desire to project and protect an image of incontestable superiority and God-like omniscient infallibility, making them appear ridiculously pathetic. They can't admit that after all the years of endorsing the non-theoretical, volume approach, maybe, just maybe, they made a mistake.

Chapter 1.5: The Evolution of Gods

Throughout history, many gods, such as those of war, peace, and love, have been worshipped. These deities reflected human traits, showing anger and demanding sacrifices. Primitive people believed natural events like lightning and hurricanes were signs of divine displeasure. To appease the gods, they offered sacrifices, often hoping to prevent further disasters.

Chapter 1.6: The Nature of Perception

Immanuel Kant, a German philosopher, explained that the nature of our mind and perceptive apparatus shapes our perception of reality. This can be compared to a computer's instruction set, which limits its computational capability. Similarly, humans have an "instruction set" in the brain that determines our knowledge and thoughts. Kant referred to the perceived and understood world as noumenal. He described the noumenal world as the world of things as they are independent of our perception, representing their essence.

Chapter 1.7: The Barrier of Perception

Our senses confine our understanding of the world, preventing us from perceiving it fully. Losing a sense, like smell, would limit our learning to the remaining senses, shaping our perception of reality.

Chapter 1.8: The Influence of Imitation

Humans naturally mimic each other, spreading language and culture. An independent voice is unattainable as we are shaped by our surroundings and experiences. My book reflects numerous influences.

Chapter 1.9: The Discovery of Cooking

Have you ever wondered how humans began cooking meat? The most plausible explanation is that during forest fires, faster animals escape, leaving only burnt meat behind. Driven by desperation and the aroma, early humans tried cooked meat, sparking their love for it and aiding in their evolution into thinking beings.

Chapter 1.10: Mastering Fire

The interest in cooked meat led humans to learn how to manage fire. Over time, humans connected fire with cooking meat. They observed that cooked meat was easier to eat and preferable in taste. To replicate this method of cooking, they learned to maintain a fire when it naturally occurred, such as from a lightning strike, until they discovered how to create fire themselves. This link between fire and cooking meat marked a significant development in human history.

Chapter 1.11: The Pursuit of Omniscience

The pursuit of knowledge and the quest for understanding the meaning of life have always been fundamental motivations for humanity. Driven by the desire to gain comprehensive insights into the world and its role within it, humans have consistently endeavored to expand their understanding. This ongoing pursuit has led to numerous discoveries and advancements, significantly shaping the trajectory of human history.

Chapter 1.12: Human Preferences, Perceptions, and the Journey to Inclusivity
Introduction: The Complex Interplay of Influences

Human preferences and perceptions are influenced by a myriad of factors, encompassing physical appearance, cultural ideals, and societal standards. These elements shape our interactions and judgments in ways that transcend the ideal of evaluating individuals based solely on character.

The Cultural Construction of Beauty

Beauty Standards and Their Origins

Beauty standards, including those related to skin color and hair type, are often culturally constructed and propagated through media, history, and social norms. The rarity of certain traits, such as blonde or red hair, can enhance their desirability simply because they are less common and often idealized. These standards are deeply embedded in societies and can influence how individuals perceive themselves and others.

Media and Historical Influences

The media plays a significant role in reinforcing beauty standards by consistently showcasing specific ideals. Historical events and cultural narratives also contribute to the perpetuation of these standards, creating a framework that shapes societal perceptions.

Embracing Diversity and Shifting Norms

Acknowledging and Appreciating Diversity

While it's important to recognize these realities, it is equally crucial to strive for a balance where diversity in all its forms is celebrated—be it physical appearance, cultural background, or individual character. Shifting societal norms and fostering greater awareness can help us move toward a more inclusive and appreciative world, even if it's not a utopia.

The Effort and Mindfulness Required

Achieving this balance is a journey that demands effort and mindfulness, both at the individual level and collectively. Embracing this complexity allows us to better understand ourselves and others, fostering a more inclusive society.

The Interplay of Structure, Logic, Function, and Control

Philosophical Reflections

In any system, be it natural, technological, or personal, there is an intricate interplay between structure, logic, function, and control. This philosophical perspective highlights how these concepts build upon each other to create cohesive systems.

Structure

A structure provides a framework, whether it is a physical building or an abstract concept. Without a solid structure, everything else falls apart. Think of it as the skeleton that supports the body.

Logic

Logic is the foundation. It ensures that everything within the structure makes sense and works cohesively. Without logic, the structure would be chaotic and inconsistent. Logic imbues the structure with meaning and purpose.

Function

Function brings the structure to life. It represents the essence of why the structure exists. Each component of the structure has a role, and together, they achieve a common goal. Functionality transforms theoretical concepts into practical applications.

Control

Control is the ability to guide and manage the structure, logic, and function to achieve desired outcomes. It is about having the power and influence to direct processes and ensure smooth operation.

Culmination of Concepts

Exclaiming "I'm in control!!!" emphasizes the culmination of these concepts. It signifies recognizing the importance of a strong foundation (logic) and a well-defined structure, understanding their functions, and ultimately, having control over them.

Human Perception and Cognitive Processing
Understanding Diverse Thoughts

Our diverse thoughts are deeply influenced by our perception of reality and the brain's unique processing mechanisms. Here are a few key points to elucidate this:

Temporal Reality and Perception

- Snippets of Reality: Our brains can only process a limited amount of information at any given moment, constantly receiving snippets of the world around us, which our brains piece together to form a coherent understanding.
- Temporal Differences: The brain processes different types of information at varying speeds. Visual information, for example, is processed almost instantaneously, while auditory information might take a bit longer, creating temporal differences in perception.

Brain Apparatus

- Sensory Processing: Our senses gather data from the environment, but each sense has its own processing time. For instance, the eyes might see something before the ears hear it.
- Individual Differences: Each person's brain has a unique structure and wiring, influenced by genetics, experiences, and

environment, resulting in different perceptions and interpretations of the same event.

Changing Environment

- Dynamic Situations: The environment is constantly changing, sometimes in the blink of an eye. Our brains must adapt quickly to these changes, leading to diverse interpretations and reactions.

Influence on Thoughts

- Cognitive Biases: These snippets and temporal differences contribute to cognitive biases, where our brain might lean towards certain interpretations based on past experiences and expectations.

- Diverse Interpretations: Because everyone's brain processes information slightly differently, two people can experience the same event but have very different perceptions and reactions.

Conclusion: Embracing the Complexity of Human Experience

"Moses, the Message, and the Mysteries"

Chapter 1.13: Moses and the Message

Moses conveyed the Bible's message using pattern recognition. It is possible to question the Bible without completely rejecting it. My perspective is derived from observation rather than religious devotion. If I were the creator and had to select a particular group of humans to deliver a message, it would be the Jews. I concur with God's choice in this regard and have found them to be highly analytical in my interactions with them. My intention is not to disprove the Bible or the claims of the Jews but to pose a small question regarding the facts presented in the Bible. Upon reading the Bible, it becomes evident from the outset that the Jews were adept problem solvers. This is demonstrated during their wars and their remarkable survival skills while escaping from bondage under Egyptian rule. They exhibit superior problem-solving and analytical skills compared to the rest of humanity and possess a better understanding of God.

Chapter 1.14: The Flaws of Fame

Regarding any moral flaws attributed to the Jews, it is their misfortune that such flaws are highlighted due to their prominence. When one is famous, one's character flaws are more exposed, and people tend to notice both the worst and the best in them. This is an inherent aspect of fame. An unknown tribe in Africa might possess similar attributes but be perceived differently due to their lack of recognition. This is where my frustration with the Bible's account of history arises. The Bible often presents an incomplete truth. It narrates epic battles and the Jews' victories and defeats but omits significant details on how these feats were accomplished or obscures the truth. The Jews devised grand battle strategies and achieved conquests, yet much of the victory is attributed to God or some supernatural story. Consequently, we receive half-truths and incomplete truths. While it is not incorrect to say they received help, the nature of this help—whether it was knowledge, supernatural intervention, or a mystical story—remains ambiguous. The Bible explicitly attributes it to the paranormal, but I infer it to be knowledge.

hapter 1.15: The Supernatural Question

Let's say we go with the supernatural or the spooky stories. It begs us to wonder why God intervenes in the Jews' affairs so much, not other races. What is so inherently special about them, not us? Moses' journal sounds much more convincing than God showing more interest in Jewish civilization, not the rest of the world. God is going to great lengths to give them special treatment and instruction on worshiping him. God ignores the rest of us. Historical evidence and observation are just for the Jews. They never spread the message to the other Semitic tribes but kept it within themselves. It wasn't Judaism that spread the message but Christianity.

Chapter 1.16: The Problem Solvers

The Jews present themselves as adept problem solvers. They have amassed wealth through their creative minds, achieving more than many others. Their ingenuity is admired by all, and they prosper wherever they go. I have observed that, regardless of geographic location, they tend to resemble or not differ significantly from the general population. This slight difference suggests that they are a mix of various other genes from the target general population. It can be inferred that intelligence is likely both genetic and social. Genetics plays a role because intelligence alone does not make one creative; rather, it is the intelligence to master one's environment that makes one truly smart.

Chapter 1.17: The Role of Intelligence

Intelligence signifies the possession of information, yet an abundance of information does not necessarily translate to productive use. Given this background, does anyone truly believe that the Jews had the Creator at their disposal, and the only thing they managed to do was witness Him perform a series of miracles? Israel faced numerous challenges, including famine, the Roman invasion, civil unrest, religious uprisings, financial ruin, and disease. Despite these adversities, none of them sought Jesus' intervention to resolve these issues, whether through violent or peaceful means. If one were to meet God, the mind that created the universe, akin to a supercomputer, the questions posed would likely extend beyond mere observation of miracles. They would encompass inquiries about the application of divine wisdom to address the pressing issues of their time.

Chapter 1.18: The Unasked Questions

"Help us turn the desert into fertile land and teach us to separate the saltwater from the ocean so we don't get thirsty." They never asked Jesus to solve complex social issues. The Jews don't get to be great thinkers by not asking the right questions, but twice they had the ears of the creators and failed to ask the hardball questions. Instead, they accepted these inspirational hymns from gods, not the equation of the Universe. In Judaism's case, I would like to think people I share similar genetic traits with would not be shallow as they are talking to the creator and only think of themselves, addressing only Jewish issues, not world issues.

Chapter 1.19: The Missed Opportunities

Jesus Christ himself said, "Give a man a fish and you feed him for a day; teach a man to fish, and you feed him for a lifetime." Instead of performing miracles, he could have imparted practical knowledge, such as how to fertilize soil for optimal crop yields or which plants could be used for medicinal purposes. He could have shared insights into effective governance or strategies to resolve conflicts peacefully. However, he chose to walk on water—an impressive feat, but not one that would save humanity. If one had the opportunity to converse with the Creator, the mind behind the universe, akin to a supercomputer, the questions posed would likely extend beyond mere observation of miracles. They would encompass inquiries about the application of divine wisdom to address the pressing issues of their time.

Chapter 1.20: The Self-Focus

The Jews primarily sought God's intervention on their behalf, rather than on behalf of others. They often acted as though they were the central focus of humanity, without considering or asking the Creator to assist others. While the Jews may have their faults, they are also known to be among the most significant charitable donors. However, in these instances, their concerns were primarily focused on themselves at that specific time. The Bible captivates readers with its powerful language, followed by these incredible and sometimes unbelievable stories. For example, the phrase "In the beginning, there was darkness" makes sense as an introduction to the narrative.

Chapter 1.21: The Spooky Stories

Regarding the details about the beginning, the notion that there was more darkness than light seems more plausible than what follows. The narrative quickly takes a nosedive into the realm of the supernatural. While there is an element of truth, the rest of the story becomes too unbelievable and forbids us from accessing the tree of knowledge. This raises the question: where are we obtaining our knowledge? The Bible presents powerful knowledge intertwined with what appears to be a fabricated supernatural story. The narrative is difficult to comprehend, and certain apologists and priests claim to understand its meaning. While some aspects are clear, the overall message is often perceived as flawed. Apologists attempt to convey their interpretations, but the delivery is often lacking. Even if we recognize in our hearts that the message is flawed, there remains a deeper feeling that there is a hidden truth if only we could see it and believe strongly enough.

Chapter 1.22: The Logistical Nightmare

Has anyone ever considered the logistical nightmare, and the mathematical computations required to sustain a nation in the desert for forty years? These were not originally desert dwellers but enslaved individuals. According to the Bible, they managed to address feeding, sanitation, watering, and health needs in an environment with limited resources, accomplishing an almost impossible task. This feat is recognized as one worthy of a god or someone blessed by the Creator. Most people rarely witnessed any miracles except for the parting of the water. Moses never explained how he accomplished this, attributing it to God, which we understand as supernatural. From a physics perspective, Moses could only have parted the sea by using two walls to separate the water or a tunnel, assuming the Red Sea was a standstill body of water like a lake. When Moses parted the sea, he and his followers must have constructed two barriers on either side, creating a pathway in the middle. Did Moses build a bridge or a tunnel?

Chapter 1.23: The Knowledge Barrier

In contemporary society, the accumulation of knowledge has increased, inversely reducing ignorance and superstition. As the world gains more knowledge, it becomes increasingly difficult to attribute everything to the supernatural. We now understand how many tasks were accomplished through practical means. The Jews, despite facing enormous challenges, continue to solve their problems without relying on the supernatural. In modern times, divine intervention appears less frequent, and the problems more complex, yet they consistently find intricate mathematical solutions.

Throughout my book, you will observe that each problem, which might seem insurmountable to other races, is tackled with persistence by the Jews. They persist as a people and even as different ethnic tribes, demonstrating remarkable resilience and problem-solving abilities.

As knowledge expands in society, superstition and ignorance tend to diminish. With a better grasp of the world, it becomes challenging to attribute events purely to supernatural forces. The Jewish people, for example, have historically overcome great challenges often through practical solutions rather than divine intervention. As divine involvement seems less prominent, they continue to find complex, often mathematical, solutions to their problems.

As knowledge expands in society, superstition and ignorance tend to diminish. With a better grasp of the world, it becomes challenging to attribute events purely to supernatural forces. The Jewish people, for example, have historically overcome great challenges often through practical solutions rather than divine intervention. As divine involvement seems less prominent, they continue to find complex, often mathematical, solutions to their problems.

Chapter 1.24: From Supernatural to Science

In contemporary times, what was once considered supernatural is now understood as science. The shift from attributing phenomena to divine intervention to recognizing them as scientific principles marks a significant evolution in human understanding. The notion that God no longer intervenes in Jewish affairs, leaving them with science, suggests that they have always possessed these tools. How else could they have addressed complex social, political, and health issues more effectively than others? It is plausible to believe that they had a head start in utilizing scientific knowledge and methodologies.

Chapter 1.25: The Head Start

Even today, the Jews are often seen as being light years ahead of the rest of the world in various fields. When an African individual asked a Chinese person why they seemed to favor Europeans over Africans, the response was that it was not personal. The Chinese explained that Europeans come to build and invest when they visit China, whereas Africans often come to extract resources and send them back home, rather than contributing to local development. This highlights a broader issue: Africans frequently assume that the discrimination they face is solely due to their pigmentation, but it is often rooted in economic factors.

If the global perception of Africans remains one of poverty, they will continue to face discrimination. This is like how society treats the homeless; rarely do we see a homeless person praised for their situation. The disparity in respect and rights between a pauper and a king is stark, and it has little to do with skin color. It reflects economic status and societal values. This is the harsh reality of the world we live in.

Chapter 1.26: The Favor of Ingenuity

Conversely, the Jews have acknowledged and appreciated China's foreigner status, recognizing their scientific advancements and ingenuity. Many Africans struggle to come to terms with this reality and often believe that the preference for Europeans over Africans is primarily based on race rather than social standing. However, it is essential to consider that economic factors play a significant role in these perceptions.

Would it be unreasonable to question the prevailing narrative and suggest that the truth is being obscured? I am skeptical of the notion that the Jews only recently acquired their knowledge. It is more plausible that they have always possessed this knowledge and have been refining it over time. This perspective challenges the idea that their success is solely due to divine intervention or supernatural abilities. Instead, it suggests a long-standing tradition of intellectual and scientific development that has contributed to their achievements.

Chapter 1.27: The Resilience of a People

Throughout the Bible, the Jews are depicted in various circumstances and predicaments with their neighbors, yet they consistently manage to overcome their challenges. Even when conquered and facing grim prospects, they persevere as resilient people. It is often portrayed that God intervenes or that something supernatural occurs to aid them, not because they are calculating the Universe's equation to escape their predicaments, but due to the presence of an all-powerful, mystical God.

This raises the question: why do the Jews receive these supernatural gifts from God while the rest of humanity does not? It is understandable for humans to accept a one-time divine intervention, but the constant meddling, adjusting, and readjusting of their affairs to the minutest details seems extraordinary. Meanwhile, the rest of humanity appears to receive no such divine assistance.

Chapter 1.28: The Journey to Canaan

The Israelites were heading for Canaan, the land of the Israelites, the Promised Land, paradise, or, as the Bible describes it, a land flowing with milk and honey. However, this land was already inhabited, by established cities, the most famous of which was Jericho. The Israelites aimed to capture Jericho, but the city was protected by impenetrable walls. How could they overcome this obstacle?

According to the Bible, God wanted to help His people. He instructed the priests to blow their trumpets, and when the Israelites heard the trumpets, they sounded like a war cry. As the story goes, the walls of Jericho collapsed, and God bestowed victory upon the Israelites over their enemies. This narrative suggests divine intervention, but sometimes legends are based on facts.

In this case, the truth may have been obscured by the supernatural elements of the story. It is plausible that the Jews created a powerful explosive device or a sonic device, possibly derived from Moses' journal, to bring down the walls of Jericho. However, instead of explaining the real story, they attributed the event to supernatural forces. To the ancients, the sound of the device would have resembled powerful trumpets heard from afar.

Chapter 1.29: The Leadership of Joshua

On the eve of battle, the aging Moses transferred the leadership of the vast nation of nomads to Joshua. As the first general of the sons of Jacob, now known as the twelve tribes of Israel, Joshua faced the daunting task of uniting the tribes, preparing them to face ruthless foes, and leading them in battles against impossible odds. After Moses' death, Joshua took command and embarked on a series of conquests.

Much of the story is experienced through the eyes of young Israelite warriors, Salmon and Tola, who personally struggle with the modern vices of the people of Canaan. Joshua, however, had his faith stretched to the limit and the will of the Israelites pushed to the brink when they finally faced the key to the conquest of Canaan: the indomitable fortress of Jericho. The narrative does not provide the battle plan or the equations they used but attributes the victory to God, presenting it as a supernatural event.

Suddenly, the Israelites were winning battle after battle, despite not being hardened warriors but rather ex-slaves. This raises questions about the true nature of their victories. Were they the result of divine intervention, or did the Israelites possess advanced knowledge and strategies that were not documented in the biblical account? The story of Jericho, with its impenetrable walls collapsing at the sound of trumpets, suggests a miraculous event. However, it is plausible that the Israelites employed sophisticated tactics or technology, such as a powerful explosive or sonic device, to achieve their victories.

This perspective challenges the traditional narrative and invites us to consider the possibility that the Israelites' success was not solely due to divine intervention but also to their ingenuity and resourcefulness. It prompts us to question the omission of these details in the biblical account and to seek a deeper understanding of the historical and scientific context behind these events.

Chapter 1.30: The Legend of Samson

Samson was a legendary Israelite warrior and judge, a member of the tribe of Dan, and a Nazirite. His immense physical strength, which he used for 20 years against the Philistines, was said to derive from his uncut hair. According to biblical accounts, God created a super-soldier for the Jews, a feat not replicated for any other people. This narrative attributes Samson's extraordinary abilities to divine intervention rather than revealing any scientific or experimental basis.

It is plausible to consider that the story of Samson might have roots in an ancient experiment aimed at creating a super-soldier, possibly derived from Moses' journal. Instead of disclosing this scientific endeavor, the narrative presents a supernatural explanation. Samson's experiment may have gone awry and was subsequently abandoned, only to be rediscovered and pursued by others, such as the Germans, much later.

This perspective challenges the traditional understanding of Samson's story and invites us to explore the possibility that his legendary strength was the result of early scientific experimentation rather than purely divine intervention. It prompts us to question the omission of these details in the biblical account and to seek a deeper understanding of the historical and scientific context behind these events.

Chapter 1.31: The Command to Destroy

The Bible contains verses about the Canaanites and the command to destroy them. When the LORD your God delivers them over to you, you shall conquer and destroy them. You shall make no covenant with them nor show mercy to them. You shall not intermarry with them, nor give your daughters to their sons or take their daughters for your sons. You shall deal with them by destroying their altars, breaking down their sacred pillars, cutting down their wooden images, and burning their carved images with fire.

This passage can be interpreted as an early instance of chemical or biological warfare, possibly derived from Moses' journal. The Israelites were instructed to cleanse themselves of the potential devastation caused by these actions. This perspective challenges the traditional narrative and invites us to consider the possibility that the Israelites' success was not solely due to divine intervention but also to their ingenuity and resourcefulness.

Chapter 1.32: The Promise of Heaven

Most Christians and Muslims believe they will indeed get to go to heaven. The Bible foretells what heaven would be like for most Christians and Muslims. Let's examine the beginning of the Bible. The Garden of Eden and the Bible reaffirm scientific theory perfectly. One of the most significant points of contention for Christians regarding science is the comparison of humans to apes. Interestingly, the Bible makes a similar comparison, albeit subliminally.

In the Book of Genesis, the Lord God commanded the man, saying, "From any tree of the garden you may eat freely; but from the tree of the knowledge of good and evil you shall not eat, for in the day that you eat from it you will surely die." This passage highlights the tension between divine command and human curiosity. The prohibition against eating from the Tree of Knowledge can be seen as a metaphor for the limits placed on human understanding and the consequences of seeking forbidden knowledge.

The story of the Garden of Eden can be interpreted as an allegory for the human journey from innocence to enlightenment. It suggests that the pursuit of knowledge, while fraught with danger, is an essential part of the human experience. The comparison to apes, both in scientific theory and subliminally in the Bible, underscores the idea that humans are part of the natural world and subject to its laws.

This perspective challenges the traditional narrative and invites us to consider the deeper meanings behind these ancient texts. It prompts us to explore the intersections between science and religion and to seek a more nuanced understanding of the human condition.

Chapter 1.33: The Consequence of Knowledge

The narrative of Adam and Eve in the Bible presents a complex interplay between divine command and human curiosity. According to the Bible, God warned Adam and Eve that they would die if they ate the fruit of the Tree of Knowledge. However, after consuming the fruit, they did not die but were instead cast out of the Garden of Eden. This raises the question: did God lie?

Consider this, Christians: without knowledge, humans are no different from animals such as dogs, cats, or chimpanzees. If Christians were stripped of all knowledge, they would be left with only the instinctual knowledge provided by the Creator during the evolutionary process. In the Garden of Eden, a man would not have been significantly different from a chimpanzee. He would not have understood the concept of grooming and would have had long hair, like any other ape. He would not have been embarrassed by his nudity, just as a gorilla is not.

The acquisition of knowledge fundamentally transformed Adam and Eve, elevating them from a state of innocence to one of awareness and self-consciousness. This transformation underscores the importance of knowledge in distinguishing humans from other animals. The story of the Garden of Eden can be seen as an allegory for the human journey from ignorance to enlightenment, highlighting the pivotal role of knowledge in shaping human identity and experience.

This perspective challenges the traditional interpretation of the biblical narrative and invites us to consider the deeper meanings behind these ancient texts. It prompts us to explore the intersections between science and religion and to seek a more nuanced understanding of the human condition.

Chapter 1.34: The Gift of Knowledge

Men were endowed with knowledge from all the animals in the Garden of Africa. However, some Christians view knowledge as a curse rather than a blessing. Christians, Muslims, and Jews often express a preference for a time when humanity was ignorant, a time when man was akin to a beast. When the woman saw that the tree was good for food, that it was a delight to the eyes, and that the tree was desirable to make one wise, she took from its fruit and ate; and she gave to her husband with her, and he ate.

What is wrong with possessing the ability to understand our world? Without knowledge, we are no different from regular apes. Knowledge elevates us, transforming us from mere animals into beings capable of understanding and shaping our environment. We are no longer just apes; we are the rising Apes, ascending through the power of knowledge.

This perspective challenges the traditional narrative that views the acquisition of knowledge as a fall from grace. Instead, it celebrates the pursuit of understanding as a fundamental aspect of human nature. The story of the Garden of Eden can be seen as an allegory for the human journey from ignorance to enlightenment, highlighting the pivotal role of knowledge in shaping human identity and experience.

By embracing knowledge, we acknowledge our potential to rise above our primal instincts and become more than mere animals. We recognize the importance of curiosity, learning, and the quest for wisdom in our ongoing evolution as a species.

Chapter 1.35: The Awakening

Then the eyes of both were opened, and they realized they were naked. They sewed fig leaves together and made themselves loin coverings. This moment signifies the awakening of human consciousness and self-awareness, setting humans apart from other animals. Our eyes are open, and we are separated from the other beasts because we possess understanding and knowledge.

Christians often attribute this newfound awareness to the devil, viewing it as the result of temptation and sin. However, this perspective overlooks the profound significance of this awakening. The acquisition of knowledge and self-awareness is a pivotal moment in the human journey, marking the transition from innocence to enlightenment.

This narrative invites us to consider the deeper meanings behind these ancient texts. It challenges us to explore the intersections between science and religion and to seek a more nuanced understanding of the human condition. By embracing knowledge and understanding, we acknowledge our potential to rise above our primal instincts and become more than mere animals. We recognize the importance of curiosity, learning, and the quest for wisdom in our ongoing evolution as a species.

Chapter 1.36: The Science Perspective

How is the scientific version of human origins that much different from the biblical account? Recall that in the beginning, humans had zero knowledge and were no different from a dog, intelligent or at least the same as a monkey or a chimpanzee. The arrival of modern science can be traced back to the scientific activities of Jesuit missionaries. According to some interpretations, the devil gave us knowledge, leading to our current understanding of human evolution.

Scientists explain human evolution as a process in which humans developed on Earth from now-extinct primates. Viewed zoologically, humans are Homo sapiens, a culture-bearing, upright-walking species that lives on the ground and first evolved in Africa about 315,000 years ago. We share a common ancestor with chimpanzees, and all apes and monkeys share a more distant relative that lived about 25 million years ago. This scientific version provides detailed insights into our origins.

The biblical narrative, on the other hand, describes the creation of humans in the Garden of Eden, where they lived in a state of innocence until they gained knowledge by eating the forbidden fruit. This act of gaining knowledge is often viewed as a fall from grace, leading to the expulsion from paradise. However, this perspective overlooks the profound significance of knowledge in shaping human identity and experience.

By comparing the scientific and biblical accounts, we can see that both narratives address the transformation of humans from a state of ignorance to one of understanding. The scientific version provides a detailed explanation of our evolutionary history, while the biblical account offers a metaphorical representation of the human journey from innocence to enlightenment.

This perspective challenges the traditional view that knowledge is a curse and invites us to celebrate the pursuit of understanding as a fundamental aspect of human nature. By embracing knowledge, we acknowledge our potential to rise above our primal instincts and become more than mere animals. We recognize the importance of curiosity, learning, and the quest for wisdom in our ongoing evolution as a species.

Chapter 1.37: The Return to Nature

Since Christians, Muslims, and Jews opposed the acquisition of knowledge from the forbidden tree and are not happy with Adam and Eve's decision, they can return to their natural state, eating and sitting in trees with no knowledge of the world around them. This would be a true animal paradise, a return to their natural habitat. Most species are truly happy when they are in their ecological systems, living in harmony with their environment. Each species has its specific place in nature, its home where it thrives.

In this context, the opposition to knowledge can be seen as a desire to return to a simpler, more instinctual way of life. Without the burden of knowledge, humans would live like other animals, guided by their instincts and natural behaviors. This perspective challenges the traditional narrative that views the acquisition of knowledge as a fall from grace. Instead, it suggests that knowledge is a fundamental aspect of human nature, one that sets us apart from other species.

By embracing knowledge, we acknowledge our potential to rise above our primal instincts and become more than mere animals. We recognize the importance of curiosity, learning, and the quest for wisdom in our ongoing evolution as a species. This perspective invites us to celebrate the pursuit of understanding as a blessing rather than a curse and to appreciate the unique role that knowledge plays in shaping human identity and experience.

Chapter 1.38: The Forbidden Knowledge

Christianity and Judaism claim that Eve ate from the forbidden fruit of knowledge but fail to specify what this forbidden knowledge entails. When infants are born, they are blank slates with zero knowledge. Humans differ from other animals, even our close relatives, by our capacity for knowledge. This forbidden knowledge gives us an edge over other animals. We can only say that humans are born with the necessary apparatus to acquire this knowledge. There has never been an infant born from the womb knowing mathematics.

The concept of forbidden knowledge in the biblical narrative can be seen as a metaphor for the human quest for understanding and the consequences that come with it. The story of Adam and Eve highlights the transformative power of knowledge and its ability to elevate humans from a state of innocence to one of awareness and self-consciousness. This transformation sets humans apart from other animals and underscores the importance of knowledge in shaping human identity and experience.

The idea that humans are born with the necessary apparatus to acquire knowledge suggests that our capacity for learning and understanding is an inherent part of our nature. This perspective challenges the traditional view that knowledge is a curse and invites us to celebrate the pursuit of understanding as a fundamental aspect of human nature. By embracing knowledge, we acknowledge our potential to rise above our primal instincts and become more than mere animals.

This perspective also prompts us to consider the deeper meanings behind ancient texts and to explore the intersections between science and religion. It invites us to seek a more nuanced understanding of the human condition and to appreciate the unique role that knowledge plays in our ongoing evolution as a species.

Chapter 1.39: The Garden of Eden Experiment

The Bible makes it quite clear on the subject, yet it is puzzling how some religious fanatics can miss it. According to the Bible, man was placed in the Garden of Eden, which was described as paradise. This controlled environment would have seemed like paradise to a beast, as all their needs were being met. The Garden of Eden can be likened to a laboratory for experimental trials. The Creator nominated humans for the experimental conscious trials, referred to as the operation knowledge tree. At that time, humans were no different from apes but exhibited slightly above-average intelligence, more than the rest of the apes or other animals.

When God noticed that Eve covered herself in His presence, He knew the new program had worked. Adam and Eve had become self-aware and were released into the general population so the new program could take effect. This scenario is akin to walking in on your dog one day and suddenly finding it ashamed of its nudity. The differentiation of humanity was part of God's plan, but the dog scenario would have been phenomenal.

Religions then emerged, telling humanity that the apparatus we were born with, which sets us apart from the rest of the beasts, is the forbidden fruit. They argue that because of this, we should follow the word of the Bible, as it is the only thing that can remove this forbidden knowledge and lead us to salvation, returning us to a life without this apparatus, back to the state of a beast.

After Adam and Eve left the Garden of Eden, they returned to Earth and had children. These modern humans, with their emotions, have not entirely mastered their newfound knowledge yet. They are spreading these so-called original sins to other humans, much like apes.

Chapter 1.40: The Evolutionary Step

In the Bible, we begin to see a slow evolutionary step toward civilizations as human apes start to self-identify, and various social groups begin to form. The other human apes would have remained like beasts, except for Adam and Eve or similar human test subjects who underwent unique experiments. This narrative aligns closely with Darwin's theory of evolution, which posits that all species are related and gradually change over time. Evolution relies on genetic variation within a population, which affects an organism's physical characteristics (phenotype). In this context, the genetic variation is that God endowed Adam and Eve with forbidden knowledge, setting them apart from other creatures.

The biblical account of Adam and Eve can be seen as an allegory for the transformative power of knowledge. By acquiring forbidden knowledge, they transitioned from a state of innocence to one of awareness and self-consciousness. This transformation marked the beginning of human civilization and the development of complex social structures. The story suggests that knowledge is a fundamental aspect of human nature, one that distinguishes us from other animals.

Darwin's theory of evolution provides a scientific framework for understanding this process. It explains how genetic variation and natural selection drive the gradual change in species over time. By comparing the biblical narrative with the scientific theory, we can see that both accounts address the transformation of humans from a state of ignorance to one of understanding.

This perspective challenges the traditional view that knowledge is a curse and invites us to celebrate the pursuit of understanding as a fundamental aspect of human nature. By embracing knowledge, we acknowledge our potential to rise above our primal instincts and become more than mere animals. We recognize the importance of curiosity, learning, and the quest for wisdom in our ongoing evolution as a species.

Chapter 1.41: The Garden of Eden and Planet of the Apes

The Garden of Eden shares a striking analogy with the movie "Planet of the Apes." In the biblical narrative, humans once dominated the Garden, living in a state of paradise until their disobedience led to their expulsion. Similarly, in "Planet of the Apes," humans once ruled the Planet until their complacency allowed the more industrious apes to overthrow them. Both stories convey a central message: human intelligence is not a fixed quality and can atrophy if taken for granted.

In the Garden of Eden, Adam and Eve's acquisition of knowledge marked the beginning of human self-awareness and the development of civilization. However, their disobedience also led to their downfall, symbolizing the potential consequences of neglecting the responsibilities that come with knowledge. This narrative highlights the importance of continually nurturing and expanding our understanding to avoid stagnation and decline.

"Planet of the Apes" presents a similar theme. The humans in the story become complacent, relying on their past achievements and failing to adapt to new challenges. As a result, the apes, who are more industrious and innovative, rise to power and overthrow the humans. This allegory underscores the idea that intelligence and progress require constant effort and vigilance. If we become complacent, we risk losing our position and being surpassed by others who are more proactive and resourceful.

Both the Garden of Eden and "Planet of the Apes" serve as cautionary tales, reminding us that the pursuit of knowledge and the application of intelligence are ongoing processes. We must remain diligent and curious, continually seeking to improve and adapt to new circumstances. By doing so, we can avoid the pitfalls of complacency and ensure that our intelligence remains a dynamic and evolving force.

Chapter 1.42: Rise of the Planet of the Apes

"Rise of the Planet of the Apes" is a 2011 American science fiction film directed by Rupert Wyatt and starring Andy Serkis, James Franco, Freida Pinto, John Lithgow, Brian Cox, Tom Felton, and David Oyelowo. Written by Rick Jaffa and Amanda Silver, it serves as 20th Century Fox's reboot of the "Planet of the Apes" series, intended to act as an origin story for a new series of films. Its premise is like the fourth film in the original series, "Conquest of the Planet of the Apes" (1972), but it is not a direct remake.

In the film, a substance designed to help the brain repair itself gives advanced intelligence to a chimpanzee named Caesar. Caesar, in turn, administers the same treatment to an animal shelter full of other great apes. This leads to an uprising against humanity after Caesar and the other super-intelligent apes suffer abuse at the hands of the shelter's staff. During the ensuing chaos, some ordinary apes escape from a zoo and join the uprising. A new society for free apes of all kinds now occupies the forests northwest of San Francisco.

The film explores themes of intelligence, freedom, and the ethical implications of scientific experimentation. It raises questions about the treatment of animals and the consequences of tampering with nature. The story of Caesar and the apes serves as a powerful allegory for the struggle for autonomy and the desire for a better life.

"Rise of the Planet of the Apes" also delves into the complexities of human-animal relationships and the potential for empathy and understanding between species. It challenges viewers to consider the moral responsibilities that come with scientific advancements and the impact of those advancements on both individuals and society as a whole.

The film's success led to the creation of a new series of "Planet of the Apes" films, further exploring the evolution of the ape society and their interactions with humans. It serves as a thought-provoking and visually stunning reimagining of a classic science fiction narrative, blending action, drama, and philosophical inquiry.

Chapter 1.43: The Paradox of Knowledge

Why are Jews, Muslims, and Christians involved with science if proper knowledge is forbidden? Shouldn't the Holy Bible be the only form of knowledge Jews should know? Everything we create in this world stems from the gains of knowledge. While religious authorities may claim that certain knowledge is forbidden, institutions like the Vatican and various Rabbis lavish on the benefits of luxurious scientific discoveries.

The wise understand themselves, while fools follow the words of others. It is essential to bring to light what is hidden in the dark—the symbols of truth. True wisdom lies not in merely seeing things but in seeing through them. Human wisdom is not about knowing and reciting sacred texts but about understanding the forces that govern human behaviors.

This perspective challenges the traditional narrative that views the acquisition of knowledge as a curse. Instead, it celebrates the pursuit of understanding as a fundamental aspect of human nature. By embracing knowledge, we acknowledge our potential to rise above our primal instincts and become more than mere animals. We recognize the importance of curiosity, learning, and the quest for wisdom in our ongoing evolution as a species.

Religious texts and scientific discoveries are not mutually exclusive. They can coexist and complement each other, providing a more comprehensive understanding of the world and our place in it. By integrating the insights from both realms, we can achieve a deeper and more nuanced perspective on the human condition.

"The Dimensions of Reality"?

Chapter 2.1: Neil deGrasse Tyson on the Multi-Dimensional Universe

We live in a four-dimensional universe, where time and place are intertwined. For example, meeting someone requires specifying both a location and a time. This concept implies that our familiar dimensions might be part of a higher-dimensional space. String theorists suggest there could be at least ten dimensions to explain all phenomena in the Universe, even if we cannot directly observe them.

We are about to embark on a dimensional journey. Picture yourself at a desk with a limited surface area. As you place pages on the desk, you eventually run out of space, having utilized the two dimensions available. To address this issue, we have developed vertical page organizers. These allow you to use the third dimension for storage, placing pages above the desk when the surface is full. If an ant were confined to the two-dimensional plane of the desk's surface, it might believe there was no more room. However, by lifting a piece of paper and placing it in the organizer, the ant would perceive the page as disappearing into an inaccessible dimension. This innovative solution allows for significantly greater storage capacity.

Consider an expanded scenario: three-dimensional rooms filled with boxes. When space runs out, an entity from the fourth dimension could intervene, moving the boxes into this higher dimension. In practical terms, this is akin to the concept depicted in Monsters Inc., where monsters utilize doors as portals through the fourth dimension to access children's closets and perform their tasks.

Imagine applying this principle to modern storage solutions. By incorporating an additional dimension, we can achieve unprecedented efficiency and capacity. For instance, envision a system where acquiring a door from a home improvement store like Home Depot enables us to create such advanced storage capabilities.

Chapter 2.2: The Portal to the Fourth Dimension

That door is a portal to a fourth dimension. You open it, place your boxes inside, close it, and they vanish. This scenario illustrates access to higher dimensions. If you hear a voice claiming ownership of your boxes from the fourth dimension, it's wise to trust it, given its superior understanding of space-time.

Does this imply that for scientists to acknowledge another dimension as having real existence, there must be evidence from beyond our current dimensions? Essentially, this suggests the need for a testimonial from an extrinsic source. In quantum physics, phenomena occur that challenge our understanding; particles appear and disappear in ways that defy traditional logic. They exhibit behaviors such as entanglement, where two particles become linked regardless of distance. Additionally, particles can traverse barriers instantaneously, seemingly faster than the speed of light, which contradicts conventional expectations.

Imagine you have a hollow sphere and live in a 2-dimensional world. As the sphere passes through your world, you will see a point appear, grow into a circle, reach its maximum size, and then shrink back to a point before disappearing. This might seem strange in two dimensions, but it makes sense in three dimensions. Similarly, some mysterious phenomena in our world could be explained by higher dimensions. Higher-dimensional physicists aim to understand these manifestations.

Chapter 2.3: The Dimensions of Reality

A point has zero dimensions. A line has one dimension: length. A square has two dimensions: height and width. A cube has three dimensions: height, width, and depth. A line is bound by two zero-dimensional points, a square by four one-dimensional lines, and a cube by six two-dimensional squares.

As we go up in dimensions, each shape's sides are one dimension less: zero-dimensional points on a line, one-dimensional lines on a square, and two-dimensional squares on a cube. A four-dimensional cube has eight sides, each a three-dimensional cube. Each side of the three-dimensional cubes is a two-dimensional square.

In four dimensions, the sides are three-dimensional surfaces. This concept leads to what is known as a hypercube or tesseract, which our brains struggle to visualize. We cannot easily imagine a volume enclosed by three-dimensional cubes. This difficulty arises because human evolution did not prepare us for such abstract concepts. As a result, we use mathematics to help us understand and represent these ideas. While we may not be able to see a hypercube directly, mathematics allows us to design models with testable consequences that can be observed and verified. This approach makes the concept acceptable within scientific inquiry.

"The Nature of Time and Gravity"?

Chapter 3.1: Time is Both Motion and Gravity

To understand that time is gravity, consider these assumptions. I propose that time and motion are inverses, have vector displacement, and are both linear and circular. We start our clock at zero motion—if everything stops moving, time stops. Assume time is circular and linear simultaneously: T1 equals linear time, and T2 equals circular time. Assume the universe is finite and spherical. A point has zero dimensions, and a ray radiates in all directions. The universe exists within a vast emptiness.

Nothingness refers to not having sufficient knowledge of what exists outside of space and time and within the Universe, which is composed of various forces such as gravity and electrical fields. Observing the Universe from an overarching perspective reveals a bright sphere within vast emptiness. By zooming in and out, one can see inside this sphere. Zooming in to Time equal to zero reveals numerous dots, while zooming out shows these dots forming molecules, complex organisms, planets, stars, solar systems, galaxies, and eventually returning to the glowing sphere, all at Time equal to zero. As time progresses from Zero forward, two patterns emerge within the sphere: the secular motion of the dots and the appearance and disappearance of certain dots. Some dots move circularly around a center of mass or nucleus as time advances, displaying two configurations at the center of Mass. Additionally, as time progresses from zero, there is an observable linear expansion of the sphere, indicative of the Universe expanding outward in all directions. The interplay between circular motion within the sphere and linear expansion results in a colliding and expanding universe. This cyclical expansion and collapse create the phenomenon known as gravity. The interaction between gravity and time equations T1 and T2 affects each atom, molecule, organism, solar system, and galaxy, from the smallest particles to the most complex structures.

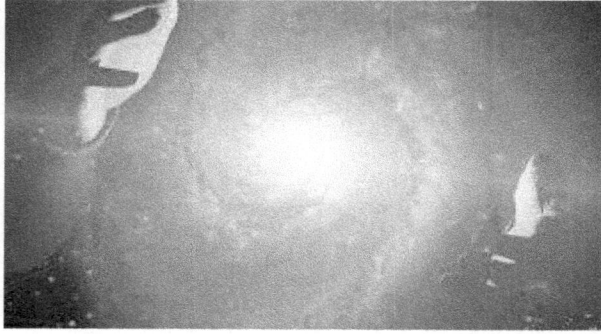

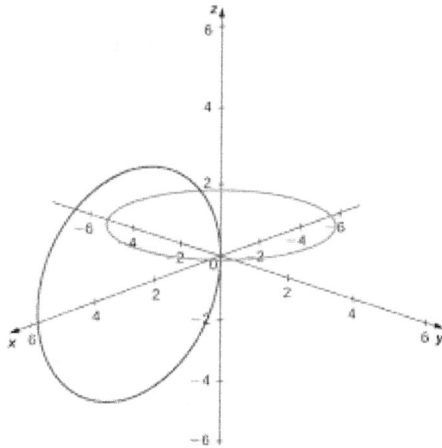

Chapter 3.2: Why is Time Not Just Linear?

To illustrate our concept, we will use an oscilloscope as an analogy. Oscilloscopes (or scopes) are instruments that test and display voltage signals as waveforms, which are visual representations of voltage variation over time. These signals are displayed on a graph showing how they change. Typically, the display monitor is circular.

Consider if a deity were observing the round Universe. If time were linear, this being would see a wave of T1, similar to how an oscilloscope displays signals, moving across the galaxy and inducing motion in a single direction throughout the Universe. This suggests that the Universe would expand directionally and elongate into a string-like form.

However, I propose that the Universe resembles an electrical field, where it grows and collapses at various points in space and time, generating gravity and maintaining its spherical nature. Additionally, I suggest that time operates in a loop, leading to the concept of a "vortex of time." In this model, our Universe functions as a vortex where time does not exist between the poles. The polarity of time transitions from temporal measurements back to motion. During these transitions, energy and matter interact, converting back and forth between forms and thus sustaining the Universe.

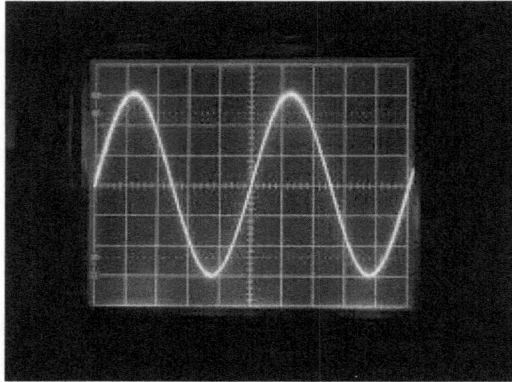

Chapter 3.3: The Big Bang and the Nature of Nothingness

The Big Bang is the leading explanation for the Universe. Before understanding it, let's define "nothingness" as being outside space and time, with no motion or displacement, and thus no time. "Nothing" means nonexistence, where reality doesn't exist, and there's no distance, time, or motion—everything appears both close and far. Imagine God in nothingness, viewing our Universe as a bright sphere of existence. By compressing this sphere to a single point, existence is squeezed into a singular reality. Alternatively, you can envision an aperture opening from reality into non-reality.

The Big Bang theory suggests the universe emerged from nothing. From a zero-dimensional point, it expanded into non-reality, forming our reality. Some theories propose reality might vanish as it grows, while others believe the universe remains unchanged by expanding into nothing. The "multiple universe theory" posits infinite universes within nothingness, each with distinct forces or possibilities, creating separate realms of existence.

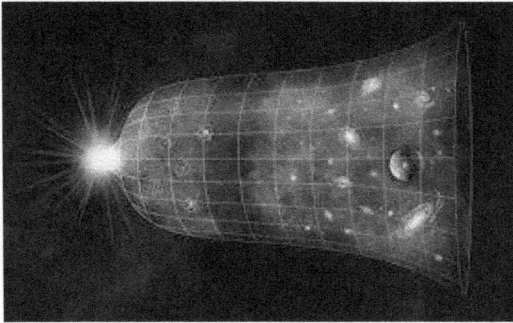

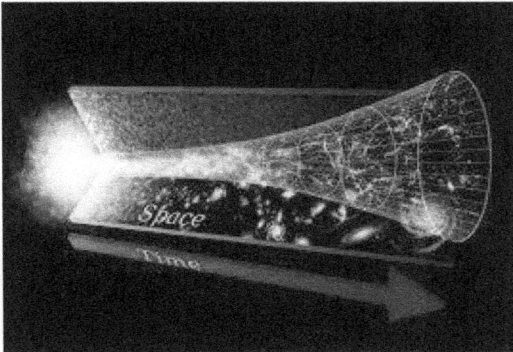

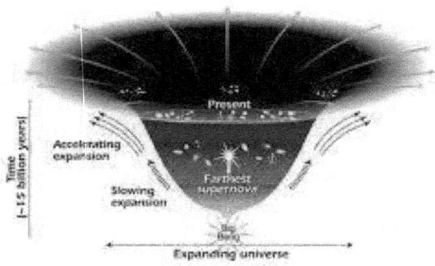

The diagram reveals changes in the rate of expansion since the universe's birth 13 billion years ago. The more shallow the curve, the faster the rate of expansion. The curve changes most notably at about 7.5 billion years ago, when objects in the universe began flying apart at a faster rate. Astronomers theorize that the faster expansion rate is due to a mysterious, dark force that is pushing galaxies apart.

Chapter 3.4: What is Gravity?

The concept of the collapse of space and time can be illustrated by bringing two balls closer together; as the spatial separation between them decreases, the temporal difference between them also diminishes. Referring to a previous discussion, "nothing" is defined as anything outside of space and time, without motion or displacement. A singularity is characterized by a function taking an infinite value, particularly in space-time contexts. Herein, we define it as an instant of creation; in the absence of a motion to separate them, all atoms would exist within one singularity, cohesively grouped. Given that space lacks dimension, a pertinent question arises: Why does everything not exist as a singular point? Space cannot be measured, having no length or motion upon examination.

Every molecule in the Universe theoretically exists simultaneously within a single instant. From a state of nothingness, a point with zero dimensions expands into a void, creating reality. While each point of reality could have coexisted as one, they remain separate. Despite having no dimension, "nothingness" results in separated instances of reality. The galaxy shows distance, with planets being spatially separated, raising questions about the nature of space. Ideally, there would be one singularity similar to the "BIG BANG". The separation of atoms or instances of reality is maintained by certain forces. Initially, all that was known consisted of a singular reality without dimensionality relative to nothingness and reality. From a state where nothing existed, something now exists. Every atom exists within a singular point, which can be understood as gravity. Atoms or instances of reality are dispersed by momentum or motion.

One hypothesis is that the Big Bang involved atoms or moments of reality repelling and attracting each other, breaking from a singularity to form the Universe and creating time, distance, and motion. As these atoms spin, they generate time. The Universe operates through the push and pull of these atoms: when close, their repulsion strengthens; when apart, their momentum weakens, potentially causing space and time to collapse around them. In the absence of motion, existence may return to a singularity.

The "pulsating universe theory" proposes that the Universe alternates between expansion and collapse due to atomic forces. These fluctuations may influence our existence and show the interconnectedness of all reality.

At the edge of the Universe, reality would violently fade in and out, making survival impossible. Light from another universe wouldn't be visible because it would have to travel through nothingness. Between universes, there's no dimension or reference point. Reality exists as a singularity; motion separates it into multiple points and shapes the known Universe's geometry. Without repulsion and attraction forces, the Universe remains a singularity. Leaving our Universe would revert you to a singularity. We exist as a hologram—appearing real but ultimately not.

Chapter 3.5: The Illusion of Space and Distance

We exist like characters in a flipbook, where space and distance are illusions. A single atom on each wall is just an instance of reality within nothingness. Space lacks dimension and relevance to time, so two instants of reality coexist in the void. If other universes exist, their locations are unknowable since everything returns to singularity. Electrons fade in and out of reality, appearing as mist. As they move further away, motion diminishes, and space and time collapse, trying to return to singularity. The concept of distance is tied to fading existence, with motion creating reality and maintaining structural integrity.

If one moment of reality created the Universe, another would create a parallel Universe. Leaving your Universe means leaving the moment that created you. You must be accepted by the other moment of reality to exist there.

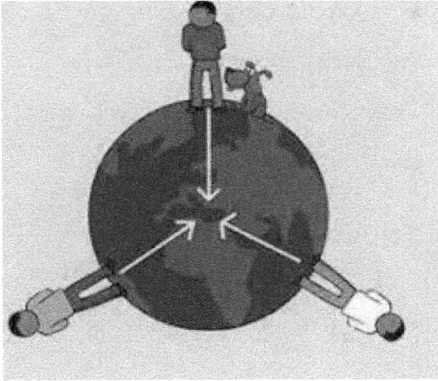

Gravity

Chapter 3.6: How Was the Holographic Universe Created?

In the void, a zero-dimensional point expands to create reality. This point, where a function reaches infinity, defines an instant of reality. All moments of reality exist as a single point or flash. Multiple seconds of reality, or atoms, are separated by movement into distinct instances, or particles, within nothingness.

Chapter 3.7: The Imaginary Dimensions of Reality

Imaginary space, time, and dimensions form around a fictitious point called a "singularity." Particles measure their proximity to this singularity with relative responses like "close but not close" or "very far." These particles avoid becoming a singularity because they represent nothingness—lacking motion, time, distance, or dimension. This creates a fictional reality as if the Universe were four-dimensional.

Chapter 3.8: The Strength of the Space-Time Continuum

The signal's strength measures the space-time continuum. Strong movement or fields indicate proximity to the singularity, while weak activity suggests a risk of returning to it.

Consider a glassblower blowing air into a tube, causing material at the other end to expand into a bubble. This bubble expands until it loses momentum, then implodes and shrinks back, trying to return to singularity. The material clumps up, repelling some parts back out, but most eventually return to singularity. This cycle continues, creating and collapsing reality from non-reality. A point with zero dimensions expands into nothingness to form reality.

Chapter 3.9: The Creation of the Holographic Universe

Reality expands into non-reality, forming a bubble that creates imaginary distance and time due to the dimension of nothingness. Revisiting the Cartesian coordinate system, each point is specified by numerical coordinates from two fixed perpendicular lines. The Universe begins as a zero-dimensional dot and expands into nothingness, defining reality. This central point, or "SINGULARITY", is an instant of reality and an imaginary center in space-time. Each instant of reality measures its closeness to singularity, forming a four-dimensional fictional Universe with length, time, and depth. This bubble of reality gives the illusion that space has dimensions, providing reference points. Inside this bubble, known universes exist, shaped by the forces acting on them, like how glass is shaped by a glassblower's force and energy.

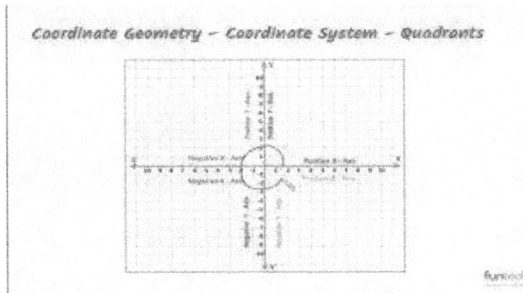

Coordinate Geometry – Coordinate System – Quadrants

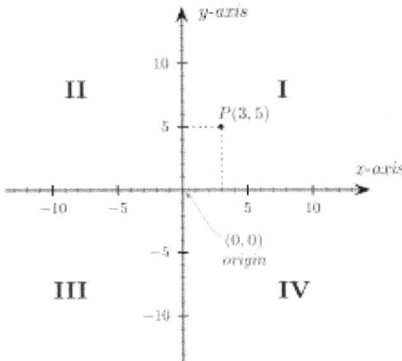

"Theoretical Teleportation and the Boundaries of Our Universe."

Chapter 4.1: Teleportation

Theoretically, a teleportation device like "The Fly" or "Star Trek Voyager" could exist, but not as shown in movies. Mixing fly DNA with human DNA is highly improbable. We will explain this later. Let's explore the concept of outside our Universe, keeping in mind that nothing exists beyond the known Universe.

Chapter 4.2: The Nature of Teleportation

Nothing has any inherent meaning; no forces exist outside our Universe. Only a "Singularity" or an "Instant of Reality" can exist beyond it, as there is no motion, distance, or time, making the cosmos dimensionless. You can't move in any direction because travel requires space and time.

In the future, we've developed a machine that can destroy reality. To teleport someone, we create a space where reality doesn't exist by removing particles to form a singularity. For instance, a human-sized vacuum tube removes all particles, turning anything inside into a singularity. If a person or fly enters, they would become a singular existence within that area.

Chapter 4.3: The Singularity and the Fabric of Reality

"Star Trek" and "The Fly" portray black holes differently. In reality, someone inside a black hole remains unchanged due to the lack of external forces. "Star Trek" fictionalizes a time limit for structural integrity. In "Singularity mode," nothing changes as no force acts on the singularity.

Creating "nothingness" in normal space-time would make it appear universally, distorting reality like a persistent cosmic stain. Distance in the universe is relative and imaginary.

One can theoretically traverse the Universe, reconstruct the structure around a gap created, introduce motion or energy, and then close it by restoring normal spatial conditions. This would allow the singularity to revert to its initial state. To alter this scenario, it must be demonstrated that incorporating "nothingness" into reality generates something new, thereby implying that reality plus zero yields reality plus an additional element.

During your journey across the Universe, time would have elapsed for you. Conversely, for an individual in a singularity state, time is non-existent; hence, they are not subject to temporal progression nor do they remain in a continuous asynchronous condition.

The concept of fly DNA mingling with human DNA is not feasible. Reality cannot be created or destroyed; only other forces can alter it. For example, a man riding a bike remains in that state unless another force acts upon him. Combining species into a hybrid would require significant changes involving time and motion. In STAR TREK, teleportation nodes are usually close together, although they could theoretically be anywhere since the void exists everywhere in the Universe. The teleportation device's non-transparent nature relates to duration—if kept open too long, it becomes visible, but if done quickly, it appears as a beam.

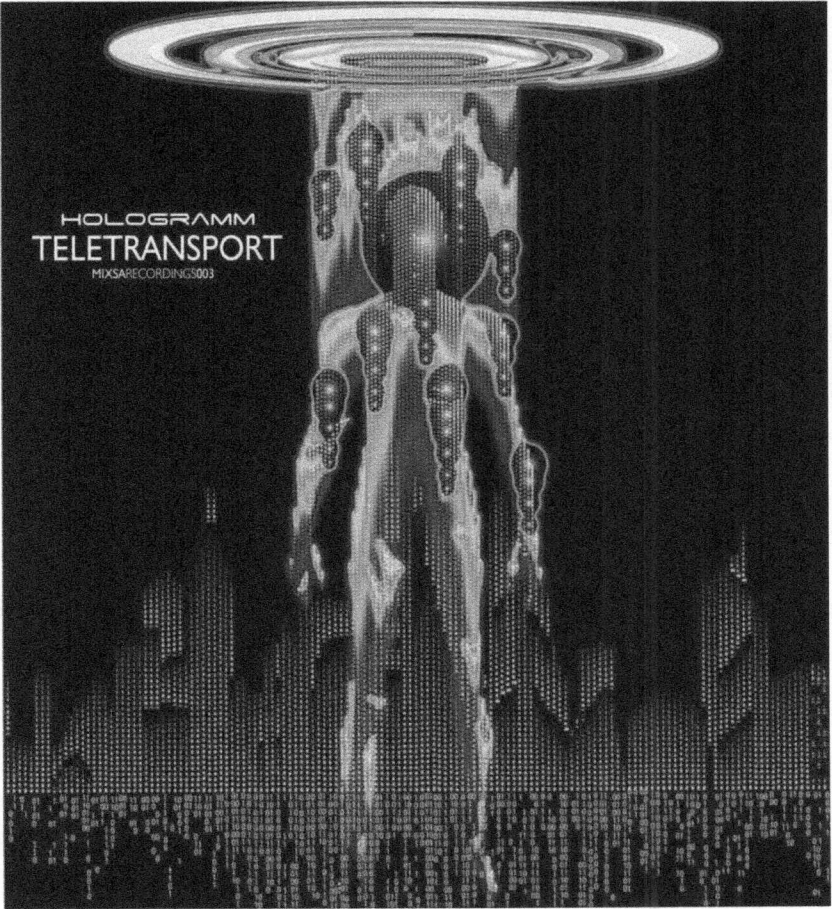

HOLOGRAMM
TELETRANSPORT
MIXSARECORDINGS003

The motion is explained mathematically.

Chapter 5.1: The Mysteries of e^t

In the vast expanse of mathematical exploration, one of the most intriguing and fundamental functions is the exponential function denoted by e^t. But what exactly is e^t? What unique properties does it possess that set it apart from other mathematical constructs?

To understand e^t, we must first delve into its defining characteristics. The function e^tt is uniquely characterized by its derivative. Remarkably, the derivative of $e^t$is itself e^t, a property that is unparalleled in the realm of functions. This self-replicating nature of its derivative is a cornerstone of its identity.

Another defining feature of e^t is the condition that when the input is zero, the function returns a value of one. Mathematically, this is expressed as $e^t = 1$. This simple yet profound condition ensures that e^tintersects the y-axis at the point (0,1), providing a foundational anchor point for the function.

As we journey through the exploration of e^t, we will uncover its profound implications and applications in various fields of study. From differential equations to complex numbers, the exponential function e^tweaves itself into the fabric of mathematical theory and real-world phenomena.

In the following chapters, we will unravel the deeper intricacies of e^t, examining its behavior, graph, and significance in both theoretical and practical contexts. Join us as we embark on this mathematical voyage, shedding light on the enigmatic and elegant nature of e^t.

$$\frac{d}{dt}e^t = e^t \qquad\qquad e^0 = 1$$

Velocity   position

e^t = position

$$e^{0.00} \qquad e^{0.44} = 1.49$$

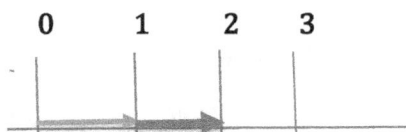

Position   velocity

Chapter 5.2: The Dynamics of Exponential Growth and Complex Motion

Imagine describing your position on a number line as a function of time, denoted by e^t. Starting at 1, your velocity, which is the derivative of your position, always equals your position. This means that the farther away you are from 0, the faster you move. This fundamental property shows how the function must grow from a specific time to a specific position, even before being able to compute e^t. You know you will be accelerating at an ever-increasing rate.

If you introduce a constant to this exponent, such as in e^{2t}, the chain rule tells us that the derivative is now two times itself, or $2e^{2t}$. At every point on the number line, rather than attaching a vector corresponding to the number itself, you first double the magnitude and then attach it. Moving so that your position is always $2e^{2t}$ is akin to moving such that your velocity is always twice your position. This scenario illustrates runaway growth, feeling increasingly out of control.

Now, consider if the constant was negative, say -0.5. Your velocity would always be -0.5 times your position vector, meaning you flip it around 180 degrees and scale its length by half. Moving so that your velocity always matches this flipped and compressed copy of the position vector, you would head in the opposite direction and slow down in a pattern of exponential decay.

But what if the constant were imaginary, denoted by i? If your position were always e^{it}, how would you move as time t progresses? The derivative of your position would now always be i times itself. Multiplying by i has the effect of rotating numbers by 90 degrees. As you might expect, things only make sense here if we start thinking beyond the number line and into the complex plane. So even before you know how to compute e^{it}, you know that for any given position at a specific time t, the velocity at that moment will be a 90-degree rotation of the current position.

$$e^{2t} = 2 \times e^{2t} \quad e^0 = 1$$

Velocity   Position

X2 →

$$e^{0.00} \qquad e^{2.05} = 2.98$$

0   1   2   3

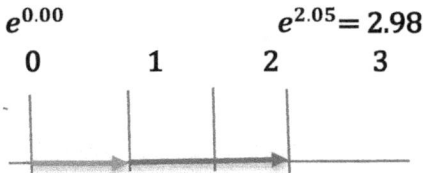

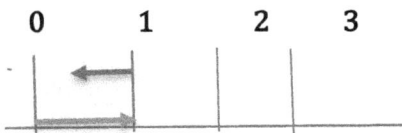

Position   velocity

B) $\dfrac{d}{dt} e^{-0.5t} = -0.5 \times e^{-0.5t} \qquad e^0 = 1$

Velocity   Position

← X (-1)

$$e^{-0.5 * 1.21} = 0.12$$

0   1   2   3

Position   velocity

c) $e^{it} \qquad e^0 = 1$

$\dfrac{d}{dt} e^{it} = i * e^{it} \qquad e^0 = 1$

Velocity

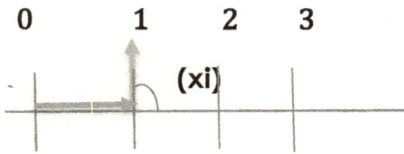

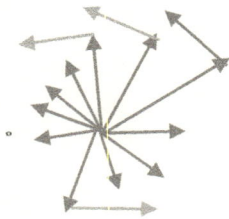

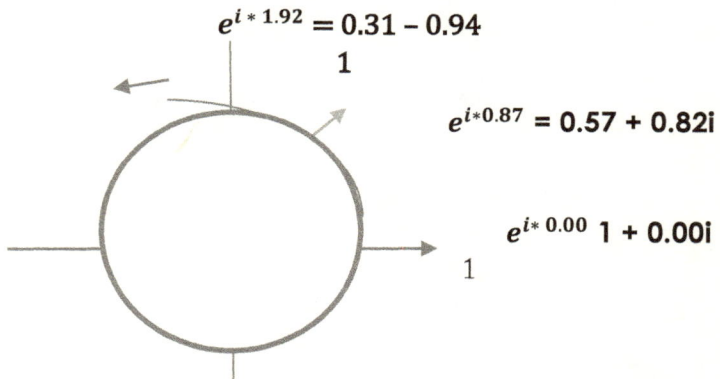

0   1   2   3

(xi)

$$e^{i*1.92} = 0.31 - 0.94$$
1

$$e^{i*0.87} = 0.57 + 0.82i$$

$$e^{i*0.00} \; 1 + 0.00i$$

1

$e^{i\pi} = -1$ e. eee... = -1
iπ times

Chapter 5.3: Visualizing Exponential Functions in the Complex Plane

As we explore the concept of exponential functions in the complex plane, we can visualize the behavior of these functions by drawing a vector field. In a vector field, we represent the vectors in a way that avoids clutter, shrinking them down for clarity.

Let's consider the function e^{it}. At time t= 0, e^{2t} equals 1. There's only one trajectory starting from this position where your velocity always matches the vector it's passing through—a 90-degree rotation of the position. This means that you move around the unit circle at a speed of one unit per second.

As you move around the circle at a speed of one unit per second, after π seconds, you have traced an angle of π radians around the circle, resulting in $e^{i\pi}$= -1. After τ seconds (where τ is equal to 2π), you have completed a full circle, bringing you back to the starting point with $e^{i\tau}$ = 1.

More generally, e^{it} equals several t radians around a circle, reflecting the continuous nature of this exponential function in the complex plane. However, one might still feel uneasy about placing an imaginary number in the exponent. This hesitation is understandable, as it introduces a layer of complexity that can seem daunting.

The expression e^t can be seen as a bit of a notational disaster, giving the number e and the idea of repeated multiplication more emphasis than they might deserve. Despite this, the behavior of exponential functions in the complex plane reveals a rich and intricate structure that is both fascinating and essential for understanding advanced mathematical concepts.

In the next chapters, we will delve deeper into the implications of these functions, exploring their applications and the insights they provide into the nature of mathematical and physical phenomena.

A Tale of Two Theories

Chapter 6.1: The Quantum Conundrum

Physics major Manoj had always suspected that the universe held more secrets than science had yet uncovered. His research led him to a profound inconsistency in energy theory between the works of two towering figures in physics: Albert Einstein and Max Planck. This revelation challenged the conventional understanding and set the stage for a new exploration of the universe's complexities.

Max Planck, the German theoretical physicist born in April 1858, revolutionized physics with his discovery of energy quanta, a breakthrough that earned him the Nobel Prize in Physics in 1918. His contributions laid the foundation for quantum theory, reshaping the landscape of modern physics.

Albert Einstein, known for his theory of relativity, also made significant strides in energy theory with his famous equation $E = mc^2$, which linked mass and energy in a transformative way.

Manoj's journey began with a simple observation: while both Planck's and Einstein's energy equations were groundbreaking, they seemed to describe the universe in fundamentally different ways. Planck's equation, which focused on energy quanta, and Einstein's mass-energy equivalence appeared to offer conflicting views of reality. This incongruity suggested that the universe was more intricate and multifaceted than previously thought.

As we delve deeper into Manoj's investigation, we will explore the implications of these equations and how they redefine our understanding of energy and the universe itself. Through this examination, we hope to uncover the hidden truths that lie at the intersection of Planck's and Einstein's groundbreaking work.

Chapter 6.2: The Quantum Puzzle

Firstly, Einstein's famous equation $E = mc^2$ tells us that mass and energy are interchangeable. This means that a given mass m can be converted into a specific amount of energy E, and vice versa. This equation is pivotal in understanding nuclear reactions, where a small amount of mass is converted into a large amount of energy.

Now, Planck's equation $E = h\nu$ relates the energy of a photon (a quantum of light) to its frequency v\nu, with h being Planck's constant. This equation is essential in the realm of quantum mechanics, describing how energy is quantized in discrete packets or quanta.

However, when we consider these two equations, we run into a bit of a puzzle. They describe energy in different contexts: Einstein's equation links mass and energy, while Planck's equation links energy to the frequency of electromagnetic waves.

To address how mass converts into radiation or light energy waves, we need to bridge these two theories. This is where the concept of mass-energy equivalence comes in. When an object with mass M loses a small bit of mass −m due to the emission of photons, the energy E released in the form of light is given by $E = (M - m)c^2$. Here, S represents the stationary object before and after the emission, meaning the total energy before and after must remain consistent due to the law of conservation of energy.

In a stationary object scenario, if mass M emits a photon, reducing its mass by −m, the energy of the emitted photon would be equivalent to the energy lost by the mass. Hence, combining both Einstein's and Planck's theories, we can say:

$$E_{photon} = mc^2 = h\nu.$$

Here, the energy lost by the mass (mc^2) is the same as the energy of the emitted photon (hν).

So, while these equations don't seamlessly exchange due to their different domains, they complement each other when we consider scenarios like the one described. They offer a unified way to understand the transformation of mass into energy and vice versa, portraying the universe as a more intricate system than initially conceived.

Physics is about reconciling these apparent contradictions and using them to reveal deeper truths about the universe.

Physics major Manoj had always suspected that the universe held more secrets than science had yet uncovered. His research led him to a profound inconsistency in energy theory between the works of two towering figures in physics: Albert Einstein and Max Planck. This revelation challenged the conventional understanding and set the stage for a new exploration of the universe's complexities.

Max Planck, the German theoretical physicist born in April 1858, revolutionized physics with his discovery of energy quanta, a breakthrough that earned him the Nobel Prize in Physics in 1918. His contributions laid the foundation for quantum theory, reshaping the landscape of modern physics.

Albert Einstein, known for his theory of relativity, also made significant strides in energy theory with his famous equation mc^2, which linked mass and energy in a transformative way.

Manoj's journey began with a simple observation: while both Planck's and Einstein's energy equations were groundbreaking, they seemed to describe the universe in fundamentally different ways. Planck's equation, which focused on energy quanta, and Einstein's mass-energy equivalence appeared to offer conflicting views of reality. This incongruity suggested that the universe was more intricate and multifaceted than previously thought.

To address how mass converts into radiation or light energy waves, we need to bridge these two theories. This is where the concept of mass-energy equivalence comes in. When an object with mass MM loses a small bit of mass –m–m due to the emission of photons, the energy EE released in the form of light is given by E= (M - m) mc^2 Here, S represents the stationary object before and after the emission, meaning the total energy before and after must remain consistent due to the law of conservation of energy.

In a stationary object scenario, if mass M emits a photon, reducing its mass by –m–m, the energy of the emitted photon would be equivalent to the energy lost by the mass. Hence, combining both Einstein's and Planck's theories, we can say:

$$E_{photon} = mc^2 = h\nu.$$

Here, the energy lost by the mass (mc^2) is the same as the energy of the emitted photon (hν).

So, while these equations don't seamlessly exchange due to their different domains, they complement each other when we consider scenarios like the one described. They offer a unified way to understand the transformation of mass into energy and vice versa, portraying the universe as a more intricate system than initially conceived.

Physics is about reconciling these apparent contradictions and using them to reveal deeper truths about the universe.

Einstein's energy equation   Planck's energy equation

$E = MC^2$ object $E = \lambda \nu$ stationary

S

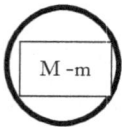   ~

$E = MC^2$ $E \neq \lambda \nu$

$$MC^2 = \lambda \nu = \frac{\lambda c}{\lambda}$$

Chapter 6.3: The Moving Paradox

As Manoj delved deeper into the intricate relationship between Einstein's and Planck's equations, he began to encounter the complexities of relativistic physics. According to Einstein's equation, $E=mc^2$, we should be able to find the decrease in mass −m concerning a stationary object. This decrease in mass corresponds to the energy emitted as radiation or light waves.

Planck's equation, E=hv, allowed Manoj to determine the energy of a wave concerning the object. Since the decreased mass –m-m is converted into a wave, we equate the energy lost by the mass to the energy of the wave, given by hv. So far, the theory has held steady.

However, the plot thickened when considering a moving reference frame. Imagine a stationary object, symbolized by SS, approaching the observer. In this scenario, the decrease in mass -m due to the emission of a photon becomes more complicated due to the relativistic effects.

According to Einstein's theory of relativity, as the object moves, its mass increases by a factor of γ, known as the Lorentz factor. This factor accounts for the effects of motion on mass and energy, given by:

The formula $\gamma = \frac{1}{\sqrt{1-\frac{v^2}{c^2}}}$ represents the Lorentz factor. It's a key component in Einstein's theory of relativity, describing how time, length, and relativistic mass change for an object moving at velocity v relative to the speed of light c.

Here's a breakdown:
- Γ (gamma) is the Lorentz factor.
- v is the velocity of the moving object.
- c is the speed of light in a vacuum (~3.00 x 10810^8 meters per second).

This factor shows how time dilation and length contraction occur as objects approach the speed of light. The closer the velocity v gets to the speed of light cc, the larger the Lorentz factor γ becomes, meaning more significant relativistic effects.

Essentially, it demonstrates how, at high velocities, time appears to slow down, and objects appear to contract in the direction of motion for an observer at rest. This plays a crucial role in understanding the relativistic energy and mass relationships in modern physics.

where v is the velocity of the object and c is the speed of light.

In the moving reference frame, the energy observed for the moving object, according to Einstein, is given by:

$$E = \gamma \, MC^2$$

Here, γM represents the relativistic mass, accounting for the variation due to the object's velocity. The energy observed by the moving object aligns with Einstein's mass-energy equivalence but is adjusted for the effects of motion.

This insight revealed a profound implication: energy and mass are not only interchangeable but also influenced by the observer's frame of reference. The transformation of mass into energy and the wave's energy all depends on whether the system is at rest or in motion. This paradox highlighted the intricate dance of relativity and quantum mechanics, suggesting that our understanding of the universe's fundamental nature is even more intricate than initially perceived.

Manoj's journey into the depths of these theories was just beginning. As he continued to explore the intersection of Einstein's and Planck's groundbreaking work, he realized that the universe's hidden truths lay in the interplay of these powerful equations, each illuminating different facets of reality.

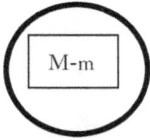

S'

M-m

Data frequency

$E = Ym C^2$

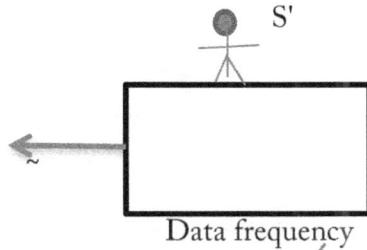

$E = v' = \dfrac{hc}{\lambda'}$ wavelength

$$\lambda' = \lambda \frac{1 - \sqrt{v/c}}{1 + \sqrt{v/c}}$$

Chapter 6.4: Reconciling Energies: Einstein vs. Planck

Manoj's revelations were profound. They revealed how the subtleties of observation could change everything we thought we knew about energy and mass.

In a moving reference frame, the energy observed for a moving object, according to Einstein, is given by:

$$E = \gamma \, MC^2$$

where γ represents the Lorentz factor, m is the mass, and cc is the speed of light. This equation shows how the energy increases with the object's velocity due to relativistic effects.

On the other hand, according to Planck's equation, the energy of a wave is given by:

$$E = h\nu$$

where h is Planck's constant, and ν is the wave's frequency.

If we consider the wavelength and frequency of the wave concerning the moving observer's frame, we use the relativistic Doppler effect to determine the observed frequency:

The equation $v' = \dfrac{v\sqrt{1-\frac{v2}{c2}}}{1-\frac{v}{c}}$ represents the relativistic Doppler effect.

Here's a breakdown of what each component means:

- v' (nu prime) is the observed frequency of the wave in the moving reference frame.
- v (nu) is the original frequency of the wave in the source's rest frame.
- v is the velocity of the source relative to the observer.
- c is the speed of light in a vacuum (~3.00 x 10810^8 meters per second).

This equation shows how the observed frequency of a wave changes due to the relative motion between the source and the observer when considering relativistic effects. It is a more advanced form of the classical Doppler effect, accounting for the high velocities involved and the effects of special relativity. The relativistic Doppler effect is crucial in understanding how light and other electromagnetic waves behave when observed from different frames of reference moving at significant fractions of the speed of light.

Given that $\lambda v = c$, we can express the energy as:

The equation $E = h\left(\dfrac{v\sqrt{1-\frac{v2}{c2}}}{1-\frac{v}{c}}\right)$ is essentially a combination of Planck's equation and the relativistic Doppler effect.

Here's the detailed breakdown:

- E is the energy of the wave.
- h is Planck's constant.
- ν (nu) is the original frequency of the wave.
- v is the velocity of the source relative to the observer.
- c is the speed of light in a vacuum.

The term inside the parentheses, ($\frac{\nu\sqrt{1-\frac{v^2}{c^2}}}{1-\frac{v}{c}}$), represents the frequency of the wave as observed in a moving reference frame. This term incorporates the relativistic Doppler effect, which accounts for how the observed frequency changes when the source of the wave is moving relative to the observer.

In this scenario, the energy observed by the moving object (according to Planck's equation) should match the energy given by Einstein's equation. This shows that the two energies, despite being derived from different principles, should ideally align when the relativistic effects are accounted for.

Manoj's exploration into these equations revealed the complex interplay between mass, energy, and observation, suggesting that the universe's true nature is a delicate balance of these fundamental forces. This deeper understanding of the universe's workings was Manoj's contribution to the ongoing quest to uncover the mysteries of the cosmos.

Chapter 5: Verifying Manoj's Relativistic Energy Transformation

$$E = \frac{\lambda}{\lambda'}$$

$$= \frac{\lambda c}{\lambda \dfrac{\sqrt{1 - v/c}}{\sqrt{1 + v/c}}}$$

$$= \frac{\lambda c}{\lambda} = \frac{1 + \sqrt{v/c}}{1 - \sqrt{v/c}} * \frac{1 + \sqrt{v/c}}{1 + \sqrt{v/c}}$$

$$= \frac{\lambda c}{\lambda} \; \frac{(1 + v/c)}{1 - \sqrt{v^2/c^2}}$$

$$MC^2 = \lambda v = \frac{\lambda c}{\lambda} = \frac{\lambda c}{\lambda} \, (1 + v/c)\, y$$

$$= MC^2 \, ((1 + v/c)\, y$$

$$= y\, MC^2 + ymvc$$

$$E' = y\, MC^2 - ymvc$$

Let's break down Manoj's steps to check for any errors in the given equation:

1. **Original Equation:**

$$E = \frac{\lambda}{\lambda'}$$

2. **Substitution:**

$$\frac{ac}{\lambda \sqrt{1 - \dfrac{v}{c}}}{1 + \sqrt{\dfrac{v}{c}}}$$

3. **Simplification:**

   This step should be simplified further.

   $$\frac{\lambda c}{\lambda} = (1 + \sqrt{\frac{v}{c}}) / (1 - \sqrt{\frac{v}{c}}) * = (1 + \sqrt{\frac{v}{c}}) / = (1 + \sqrt{\frac{v}{c}})$$

4. **Further simplification:**

   The equation $\frac{\lambda c}{\lambda} = \left(\dfrac{1 + \sqrt{\frac{v}{c}}}{1 - \sqrt{v^2/c^2}} \right)$ appears to contain a combination of elements related to relativistic Doppler shifts and wavelengths.

Here's what each component signifies:

- λ (lambda) represents the wavelength of the wave.
- C (capital C) likely stands for the speed of light.
- v is the velocity of the source relative to the observer.
- c (lowercase c) is the speed of light in a vacuum (~3.00 x 10810^8 meters per second).

The term on the left-hand side $\frac{\lambda c}{\lambda}$ simplifies to C, suggesting a relationship between the speed of light and other parameters.

The right-hand side ($\dfrac{1 + \sqrt{\frac{v}{c}}}{1 - \sqrt{v^2/c^2}}$) seems to imply a modification of the relativistic Doppler effect. This combination of terms accounts for the effects of the source's velocity on the observed wavelength.

However, there is a discrepancy in the right-hand side involving the square root, which might need a closer examination to ensure that the expression aligns correctly with relativistic principles.

5. **Equation Expansion:**

$$MC^2 = \lambda v = \frac{\lambda c}{\lambda} = \frac{\lambda c}{\lambda}\left(1 + \sqrt{\frac{v}{c}}\right)\lambda$$

6. **Additional Simplification:**

$$MC^2\left(1 + \sqrt{\frac{v}{c}}\right)\lambda = \gamma\, MC^2 + \gamma mvc$$

7. **Final Form:**

$$E' = \gamma\, MC^2 - \gamma mvc$$

Let's summarize the findings:

- The original substitution seems correct.
- When simplifying the terms, it appears that there may be an inconsistency in the simplification step.
- Double-checking the steps might be necessary to ensure that the relativistic Doppler factor and other terms are handled appropriately.

Chapter 6: Expanding on Manoj's Computational Findings

Manoj noticed a discrepancy: Planck's Energy appeared greater than Einstein's Energy under certain conditions. Given that the mass M is converted into the wave of light, the energy of the mass MM and the radiant wave should theoretically be the same. However, he found the following:

1. **Energy of the Mass M:**

$$E = \gamma\, MC^2$$

2. **Energy of the Wave:**

$$E = \gamma\, MC^2 + \gamma Mvc$$

Here, γ is the Lorentz factor, M is the mass, c is the speed of light, and v is the velocity of the source. The energy discrepancy raised the question: where is the excess energy coming from, and why does it appear?

Steps to Check for Errors

1. Reexamine Einstein's and Planck's Equations:

    Einstein's Equation:

    $$E = \gamma\, MC^2$$

    Planck's Equation:

    $$E = h\nu$$

For a moving object, incorporating the relativistic Doppler effect:

$$v' = \frac{v\sqrt{1 - \frac{v^2}{c^2}}}{1 - \frac{v}{c}}$$

2. Energy Conversion: When considering the energy conversion for a wave, we need to equate the energy from Einstein's equation to that from Planck's equation, adjusting for relativistic effects.

3. In a Moving Reference Frame:
    *   For an object moving towards the observer, the energy observed using Einstein's equation is:

    $$E = \gamma\, MC^2$$

Using Planck's equation and the Doppler-shifted frequency:

$$h \left(\frac{v\sqrt{1-\frac{v^2}{c^2}}}{1-\frac{v}{c}} \right)$$

Simplified, this should align with:

$$E = \gamma \, MC^2 + \gamma Mvc$$

However, if we consider an object moving away:

$$E' = \gamma \, MC^2 - \gamma Mvc$$

Here, Planck's energy appears less than Einstein's energy.

Possible Explanations for Discrepancies

- Relativistic Effects: The discrepancies could stem from the complex interplay of relativistic effects. When objects move close to the speed of light, the energy observed changes due to both time dilation and length contraction.
- Energy Conservation: The principle of conservation of energy must still hold. The apparent excess energy might be accounted for by considering additional relativistic corrections.

Manoj's findings invite further questions about our understanding of mass and energy transformation in the universe. Are there hidden aspects or corrections we haven't fully considered? Could there be additional factors influencing these equations?

Solving Mr. Manoj's Mystery

Chapter 7.1: Bridging Theory and Reality

Manoj's discovery of inconsistencies in energy theory between the works of Einstein and Planck was just the beginning. To address his questions, we need to explore his findings deeply and see how they align with our broader understanding of the universe.

Using my Universe model, I will attempt to solve Mr. Manoj's questions, shedding light on the nuances he has uncovered. Manoj's assertion points to a significant truth in the world of science: theory and application often differ. While theories make various assumptions to explain phenomena and concepts, real-life conditions are unique and do not always conform to these assumptions. This dichotomy forms the backbone of all learning procedures, illustrating the ongoing dance between theoretical understanding and practical application.

Manoj's central concern revolves around the discrepancy between the energy equations of Einstein and Planck when considering a moving reference frame. Let's delve into his observations and expand upon them.

Theoretical Framework

In theory, Einstein's famous equation $E = MC^2$ provides a straightforward relationship between mass and energy. It implies that mass can be converted into energy, and the amount of energy produced is equal to the mass multiplied by the square of the speed of light. Planck's equation $E=h\nu$, on the other hand, relates the energy of a photon to its frequency. These equations form the cornerstone of modern physics, each describing a different aspect of the universe.

Practical Application

However, when we apply these equations in real-life scenarios, especially involving relativistic speeds, things become more complex. Consider a stationary object emitting a photon. According to Einstein, the energy of the mass M is given by γMC^2, where γ is the Lorentz factor. This equation holds in a stationary frame of reference.

Now, when the object is moving, the energy of the wave can be expressed as:

$$E = \gamma MC^2 + \gamma Mvc$$

This additional term γMvc introduces the velocity of the moving object, accounting for the relativistic effects. Here, the discrepancy arises. Manoj observes that, in theory, the energy of the mass M and the energy of the resultant wave should be the same. However, due to the relativistic effects, the energy observed in a moving frame appears different. Where is this excess energy coming from? This question leads to deeper insights.

Addressing the Discrepancies

Relativistic Effects: The additional term γMvc can be attributed to the effects of motion at relativistic speeds. The energy observed changes due to the velocity of the source, introducing complexities not apparent in a stationary frame.

Energy Conservation: Despite these discrepancies, the principle of energy conservation remains intact. The apparent excess energy might be accounted for by considering the complete relativistic framework, including time dilation and length contraction.

Moving Away Scenario: When the object is moving away, the energy equation transforms to:

$$E' = \gamma MC^2 - \gamma Mvc$$

In this case, Planck's energy appears less than Einstein's energy. This again highlights the influence of relative motion on observed energy, suggesting that the universe's true nature is far more intricate than previously thought.

The Dichotomy of Theory and Practice

Chapter 7.1: The depths of Einstein's and Planck's theories

Manoj's observation underscores the eternal dichotomy between theory and practice. Theories, with their simplified assumptions, provide a framework for understanding. However, real-life conditions, with their unique and complex nature, often challenge these theoretical models. This dichotomy is not a flaw but a driving force in the pursuit of knowledge, pushing scientists to refine and expand their understanding continually.

As we explore these complexities, we realize that each discovery leads to more questions, driving the relentless quest for knowledge. Manoj's journey into the depths of Einstein's and Planck's theories is a testament to this ongoing pursuit, revealing the hidden truths of our universe.

In the following chapters, we will continue to unravel the mysteries of the cosmos, bridging the gap between theory and practice and uncovering the profound insights that lie beyond the equations.

This deeper dive into Manoj's findings not only addresses his questions but also highlights the beautiful complexity of the universe. Let us continue this journey, exploring the unknown and pushing the boundaries of our understanding.

Mr. Manjor says that converting Mass into Energy or Energy back to Mass sounds good and works, but when you apply a signal, positive or negative Energy is different from one state to another. Mr. Manjor asks why this is occurring. I will give an analogy; an electrical circuit has power; we expect to get in theory and what we get are two different things.

Real power is the power consumed due to the resistive load, and apparent power is the power the grid must withstand. The unit of absolute power is the watt, while the apparent power unit is VA (Volt Ampere)

The total power flowing is the "apparent power," We measured it as the voltage and current (V * I). For example, if 208 volts and five amps are measured – the apparent power is 1040VA (VA means volt-amps – the measurement unit of apparent power).

The equipment consumes true power to do practical work in an AC circuit. It is distinguished from apparent power by eliminating the reactive power component that may be present.

Mr. Manoj Question is a fair one. Why is the energy different? Is there something consuming the Energy or adding Energy to the equations? Using my model, I will attempt to solve this riddle with the expanding universe model complex to explain why it's happening. I retraced Mr. Manoj's logic, equations, and findings to support my theory, so I used it.

As you can see, the calculation is very tedious. Yet, Mr. Manoj brought up a very valid question. We see the theory and apply it and work, but it behaves differently, as shown in the calculation above, self-evident when you input a signal.

Chapter 7.2: The Energy Paradox

Mr. Manoj asserts that while the concept of converting mass into energy, or vice versa, works in theory, practical applications reveal discrepancies. When we apply a signal, the energy observed can differ from one state to another. This raises the question: why does this occur?

To illustrate, consider an analogy with electrical circuits. In theory, we expect certain power values, but in practice, real power (consumed due to resistive load) and apparent power (which the grid must handle) can differ. Real power is measured in watts, while apparent power is measured in volt-amperes (VA).

Apparent power (V×IV \times I) represents the total power flow, such as measuring 208 volts and 5 amps to yield 1040 VA. However, true power, consumed for practical work, excludes any reactive power component.

Mr. Manoj's question is valid: why is the energy different? Is there something consuming or adding energy? Using my Universe model, I'll attempt to unravel this riddle.

Addressing the Discrepancies

Relativistic Effects: The differences can be attributed to the effects of relativistic speeds. The energy observed changes due to the source's velocity.

Energy Conservation: Despite discrepancies, the conservation of energy principle remains intact. The perceived excess or deficit might be accounted for by considering the complete relativistic framework.

For example, when an object is moving towards the observer, the energy of the wave includes an additional term due to the object's velocity. Conversely, if the object is moving away, the energy appears reduced. These variations suggest a more intricate universe than initially thought.

Conclusion

Mr. Manoj's inquiry highlights the perpetual challenge of bridging theory and practice. Theories provide a framework for understanding, but real-life applications reveal complexities not captured by simplified models. This dichotomy drives the relentless pursuit of knowledge, pushing us to refine our understanding continually.

Through detailed calculations and re-examination of foundational equations, we begin to uncover the hidden truths of our universe. Let us continue exploring these mysteries, bridging the gap between theoretical predictions and practical observations.

We will follow the logic of my model to solve this mystery. We began to define reality and follow the logic of my model, and you will see how easy it is to solve this mystery. Let's say reality is a change in motion in time. Something changes its motion in time, saying that something is real. The various forms of reality are dark matter, electrons, protons, neutrons, and knowledge energy; we define these states of matter based on their motion or energy level. We can say that Energy and matter are the same things but differently. It is safe to assume that motion is Energy and is related. Each state of matter or Energy depends on its state of matter. The higher the Energy or motion, the closer they become to Energy. I hypothesize that all the basics are reality but have different energy levels in my theory. The electron has an energy level; dark matter has an energy level, and all others.

For example, things are the same but different at the same time. Let's say you are a mixed person of European and African. You can either identify as European with African heritage or African of European heritage. Either way, you are right. It's just a matter of perspective, which races are closer. I like this analogy because the one-drop rule nullifies a person from being a European of African descent; if you are European and have one drop of black blood, you are black.

Since, in my hypothesis, they are all instants of reality, and the difference is their energy level, it stands to reason that moving from one state to another requires adding or subtracting Energy. Since wave reality is faster or has more Energy or a higher level, it needs to gain Energy in that state.

Chapter 7.3: Defining Reality Through Motion

We will follow the logic of my model to solve this mystery. By defining reality and adhering to the logic of my model, you will see how easy it is to unravel this conundrum. Let's say reality is a change in motion in time. When something changes its motion in time, we acknowledge its existence.

Various forms of reality include dark matter, electrons, protons, neutrons, and knowledge energy. We define these states of matter based on their motion or energy level. We can argue that energy and matter are fundamentally the same, expressed differently. It is safe to assume that motion is synonymous with energy and that they are interconnected. Each state of matter or energy depends on its energy level. The higher the energy or motion, the closer it becomes to pure energy.

I hypothesize that all basic elements of reality possess different energy levels. For instance, electrons, dark matter, and other forms have distinct energy levels. Things may appear the same yet differ simultaneously. Consider a person of mixed European and African descent. One could identify as European with African heritage or as African with European heritage. Both perspectives are valid. It's a matter of perspective, much like the "one-drop rule" that historically classified individuals with any African ancestry as Black.

In my hypothesis, each state of matter or energy is an instance of reality, differentiated by their energy levels. Moving from one state to another involves adding or subtracting energy. For example, wave reality, which moves faster or possesses more energy, requires an increase in energy to reach that state.

In the context of Mr. Manoj's query, understanding these principles can shed light on the discrepancy between observed energies. If the energy levels of different states are considered, the varying energy requirements make sense. By following this model, we can better comprehend the nuances of energy and matter, providing clarity to the mysteries of the universe.

Chapter 8.1: Fluid Dynamics and Understanding It from the Quantum Level

In this chapter, we delve into the fascinating world of fluid dynamics, exploring both familiar phenomena like lift and drag, and less familiar ones like vortex shedding and the interplay between pressure and velocity. Our journey begins at the subatomic particle level, the foundation of molecular interactions.

To reduce the complexity of this topic, we will employ different approaches. One such approach involves averaging the molecules to establish the fluid as a continuous field. This continuum is then discretized spatially and temporally, whatever that may entail. Finally, we wrap it all up with proper boundary treatment. To lay the mathematical foundation, we must immerse ourselves in fields like computational fluid dynamics (CFD).

CFD can initially seem daunting, much like an unsettling force. However, the key to mastering it lies in breaking down its complexity into a few core ideas and eliminating sensory overload. By focusing on the microscopic foundation, we justify the macroscopic implementations of fluid dynamics. My goal in introducing this topic is to help you grasp the microscopic perspectives of fluid dynamics.

Most of us have a general concept of how fluids work. Some might view fluid dynamics as a field of discrete colliding particles, while others see it as a continuous flowing medium where forces act on objects at a distance. Our first task is to align our perspectives on fluids. Both viewpoints are valid and used more or less depending on the context.

Understanding fluid mechanics begins with distilling information to its essence. It's impractical to process all raw data simultaneously due to the sheer number of molecules in a fluid object. Hence, we focus on statistically significant information while discarding the rest, such as the rotation of individual molecules.

This chapter sets the stage for our journey through fluid dynamics, emphasizing the importance of simplifying complex information to uncover the fundamental principles governing fluid behavior. Through this approach, we aim to bridge the microscopic and macroscopic worlds, providing a comprehensive understanding of fluid dynamics from the quantum level.

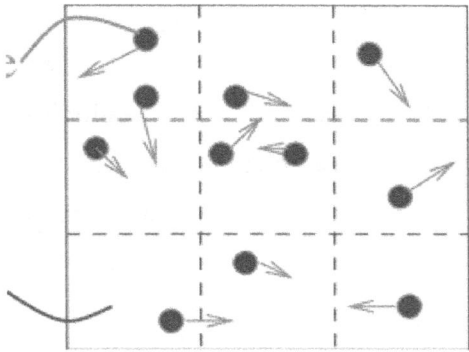

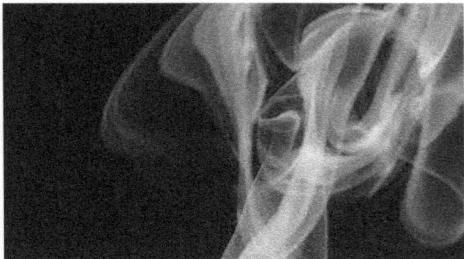

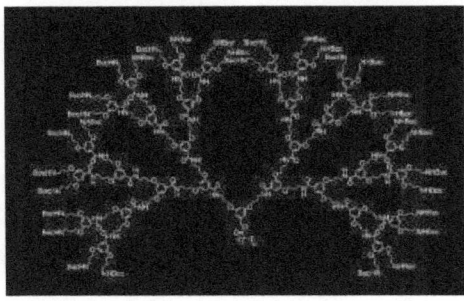

Chapter 8.2: Analyzing Molecular Impact on Fluid Dynamics

Consider the rotating of a molecule. Does this rotation make a significant contribution to phenomena like lift if it were different? Or think about the exact location of each molecule. In practice, what truly matters is the average behavior of a small sample or handful of molecules. Imagine the vast amount of charts and data required for such an exhaustive task.

Fortunately, we are primarily interested in quantities of a global character, such as the net lift force for a given instance of reality. This approach allows us to focus on the broader picture, examining the overall effects rather than getting lost in the minutiae of individual molecular behaviors.

By concentrating on these global quantities, we can simplify our analysis and derive meaningful insights into fluid dynamics. This perspective enables us to bridge the gap between the microscopic interactions and macroscopic outcomes, providing a clearer understanding of how fluid dynamics operate at both levels.

Aerodynamic Lift – Explained by Bernoulli's Conservation of Energy Law

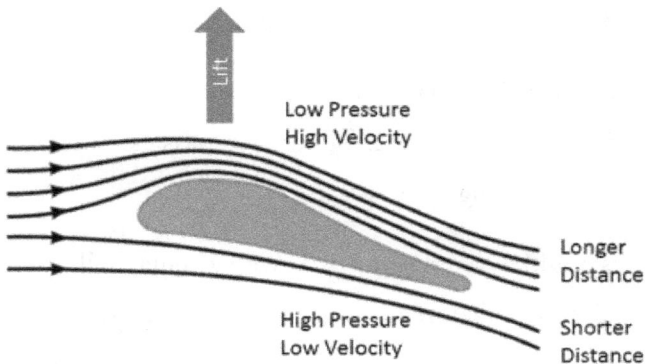

Lift

Low Pressure
High Velocity

Longer
Distance

High Pressure
Low Velocity

Shorter
Distance

Also known as the "Longer Path" or "Equal Transit" Theory

Chapter 8.3: Mathematical Relevance in Fluid Dynamics

Mathematically, how do we determine whether certain aspects are relevant or not? One might assume it's a straightforward, logical process, but it's more complex than it seems. Our approach involves several layers or modules, each designed to systematically break down the problem into manageable parts.

Firstly, we consider the significance of the variables involved. Not all variables hold the same weight in every scenario, and identifying which ones are crucial requires careful analysis. This step ensures we focus on the most impactful factors, eliminating unnecessary complexity.

Secondly, we employ statistical methods to average out the behavior of small samples. By doing so, we gain a clearer picture of the overall trends and patterns, which are often more informative than individual data points.

Thirdly, we utilize computational models to simulate various conditions and predict outcomes. These models help us understand the dynamic behavior of fluids under different circumstances, providing insights that are difficult to obtain through direct observation alone.

Finally, we integrate boundary conditions to refine our models further. Properly defining the boundaries of our system is essential for accurate simulations and meaningful results.

Through this multi-layered approach, we systematically evaluate the relevance of different factors in fluid dynamics. By breaking down the problem and focusing on the most significant elements, we can develop a deeper understanding and more accurate predictions, bridging the gap between theory and practical application.

In the following chapters, we will apply this approach to various fluid dynamic scenarios, exploring how these principles manifest in real-world situations and uncovering the hidden complexities of fluid behavior.

Chapter 8.4: Building Blocks of Quantum Mechanics

In any comprehensive module, it is the essential building blocks that give structure and clarity. These foundational tools will be vital for our explanations and offer a solid perspective. Before we can build from the ground up, a good starting point is necessary, and quantum mechanics provides just that.

To grasp its complexities, we need to break quantum mechanics into various layers of abstraction, addressing the underlying issues within each subcomponent. What fundamental problem does quantum mechanics aim to solve or approach? This problem is so intricate that even simulating a small fluid

The system at the quantum level poses significant challenges.

Understanding these challenges requires us to delve into the fundamental properties of nature, which ultimately boil down to the way we think about and conduct measurements. Quantum mechanics addresses how particles behave and interact at the smallest scales, where traditional physics no longer applies.

By examining these foundational elements, we lay the groundwork for understanding fluid dynamics at a quantum level. This approach allows us to appreciate the microscopic interactions that dictate macroscopic behaviors, bridging the gap between the seemingly disparate worlds of quantum mechanics and fluid dynamics.

In the upcoming chapters, we will explore these layers in detail, uncovering the nuanced complexities that define quantum mechanics and its application to understanding the fundamental nature of fluids.

Through this journey, we aim to provide clarity and insight into one of the most challenging and fascinating areas of modern science.

Chapter 8.5: Tracking the Unseen

Tracking macroscopic objects and their paths or trajectories is straightforward. One can visually observe these objects or use electronic tracking devices to monitor their movements. However, the challenge intensifies when we shift our focus to microscopic entities, such as elementary particles.

How do you track the path of something as minuscule as an electron orbiting the nucleus of a hydrogen atom? This task is far more complex. Traditional tracking methods become ineffective due to the scale and behavior of these tiny particles.

At the quantum level, particles like electrons do not follow well-defined paths as larger objects do. Instead, they exhibit wave-particle duality, existing in a probabilistic state until measured. This makes tracking their exact trajectory virtually impossible. Instead, we rely on statistical methods and probability distributions to predict their behavior.

Quantum mechanics provides the framework for understanding and tracking these subatomic particles. Through mathematical models and experimental techniques, we can infer the likely locations and paths of particles. Tools such as the Heisenberg Uncertainty Principle highlight the limitations of precisely measuring both the position and momentum of a particle simultaneously.

In this chapter, we explore the methods and principles used to track the movements of elementary particles. By understanding these techniques, we gain insight into the fundamental nature of the microscopic world, bridging the gap between the seen and unseen, and unveiling the intricate dance of particles at the quantum level.

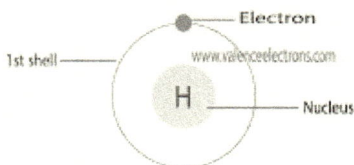

Chapter 8.6: Rethinking Observation

To truly understand the microscopic world, we must rethink what "looking" means. Let's build an experiment to illustrate this concept.

Imagine we use a laser to shoot photons towards an atom placed within an electric field. When a photon strikes the atom, it might kick an electron out of the electrostatic potential well surrounding the atom. This interaction causes the electron to be ejected and subsequently accelerated by the electric field.

As the electron moves to the right side, its presence and influence can be detected. This process of observation highlights the intricate nature of measuring and tracking subatomic particles. By rethinking observation, we gain a deeper understanding of the dynamics at play and how these tiny interactions shape the behavior of matter at the quantum level.

In the following chapters, we will continue to explore experimental techniques and theoretical frameworks that provide insight into the fundamental properties of nature. Through these investigations, we aim to bridge the gap between what is seen and what is understood, unveiling the complexities of the universe one experiment at a time.

Electron Detector

Location sensitive

Electric Field Potential

Iterating downwards

Chapter 8.7: The Challenge of Quantum Observation

If our experiment is designed correctly, it becomes possible to locate the original position of an electron within an atom. Imagine being able to pinpoint where an electron is in an atom. However, in doing so, we significantly alter the electron's future trajectory. In this context, "looking" inherently means "interacting."

The specimens and the measuring devices—electrons and protons—operate at an energy scale far beyond any macroscopic composition. Consequently, while measuring, we inevitably exert influence. Even if the experiment is non-destructive to the atom, we must interact with the electron in some form to detect it.

This interaction is intrinsic to the nature of quantum measurement. As we try to observe such minute particles, our methods of detection inevitably disturb their natural state. This is a fundamental principle of quantum mechanics known as the observer effect. It states that the act of observation alters the state of the particle being observed.

Thus, tracking an undisturbed trajectory of an electron, or any similarly small particle, appears nearly impossible. Every measurement imposes an influence, making it challenging to capture the true undisturbed path of such particles.

Understanding and acknowledging these limitations is crucial for advancing our comprehension of quantum mechanics. By rethinking our approach to observation, we can better appreciate the complexities involved in tracking and measuring subatomic particles, and continue to push the boundaries of our knowledge in the quantum realm.

Position   Momentum
                                  (Mass x velocity)

    Standard deviation

⬭   ○   Possible

    Uncertainty relation

Chapter 8.8: The Limitations of Quantum Measurement

Unfortunately, a fundamental limit prevents us from determining the exact location and momentum of particles, regardless of the Heisenberg uncertainty principle's measurement process. Theoretically, the trajectories of such small entities are negligible. However, averaging many observations to yield meaningful information retains its significance from a probabilistic perspective. In the end, how likely is it to detect an electron in a particular region of the atom?

The Heisenberg uncertainty principle in quantum mechanics posits that there is a fundamental limit to the accuracy with which specific pairs of physical quantities of a particle, such as position (xx) and momentum (pp), can be predicted from initial conditions.

The formula is as follows:

$$\Delta x \Delta p \geq \frac{\hbar}{\pi}$$

Where:

- Δx = uncertainty in position
- Δp = uncertainty in momentum
- $\hbar$ = Planck's constant divided by 2π
- π = pi

Depending on properties such as the electron's energy or the strength of the electric field, these probability distributions can assume various shapes. The situation becomes even more complicated if multiple electrons are present around an atom, as the motion of one electron is correlated with the others. Therefore, an accurate multi-electron probability distribution cannot simply be a composition of single-electron shapes.

This chapter delves into the intricate nature of quantum measurement, highlighting the challenges posed by the Heisenberg uncertainty principle. By understanding these limitations, we can better appreciate the probabilistic nature of quantum mechanics, and the complexities involved in measuring and predicting the behavior of subatomic particles. This exploration bridges the gap between theoretical predictions and practical observations, deepening our comprehension of the quantum world.

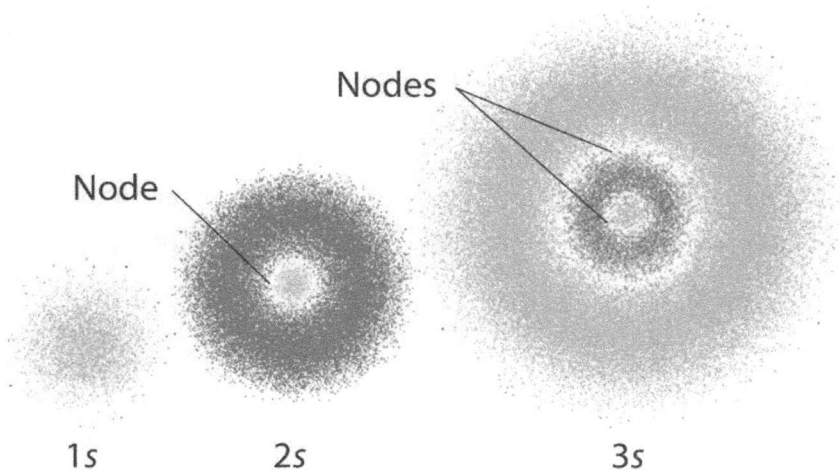

Node

Nodes

1s   2s   3s

Although we can approximate them as a starting point, the result is always a probability distribution.

Chapter 8.9: The Dual Nature of Elementary Particles

An intriguing aspect of small elementary particles is their dual behavior. They behave not only like particles, meaning they exist in discrete quantities, and once measured, their impact is observed at specific locations. But they also behave like waves, allowing them to go around corners, superimpose, diffract, and interfere when not observed.

To make sense of these probabilities and wave-like properties, science has developed a mathematical framework that reflects the indeterminate nature observed in experiments. At its core, this framework is known as quantum mechanics.

Quantum mechanics relies on two fundamental mathematical objects:

1. **Wave Function:** This function contains information about the state of all considered particles.
2. **Schrödinger Equation:** An evolution equation that operates on wave functions, it tells us how these functions change over time.

The Schrödinger equation is a linear partial differential equation that governs the wave function of a quantum-mechanical system. It is a crucial result in quantum mechanics, and its discovery marked a significant milestone in the field. This equation allows scientists to predict the behavior of particles over time, considering their wave-like and particle-like properties.

Understanding the dual nature of elementary particles and the mathematical tools developed to describe them provides deep insights into the behavior of matter at the smallest scales. This knowledge bridges the gap between theoretical predictions and experimental observations, advancing our comprehension of the quantum realm.

Quantum Mechanics

Wave Function   Evolution equation

$\omega =$ $+ i$ $\partial\partial\ \psi = F(\psi)$

Real   Imaginary

Probability density

2   Probability Distribution

Wave   Waves go around corners.

Chapter 8.10: From Wave Functions to Probabilities

From the new wave function, we can derive probabilities, which tell us how likely it is to detect a particle in a specific region of space. This is the same kind of probability we've discussed before. Essentially, we have a model that reflects what we observe in our measurements. The exact form of the evolution equation is less important than understanding the underlying concept. Our primary concern is how computationally expensive it would be to perform such calculations for a given volume of fluid. For this, we need to know the time required for the evolution operation.

For example, the wave function assigns a complex number to each point in a two-dimensional space:

$$= +0.00 - 0.00i$$

The evolution equation, governed by the Hamiltonian, takes these numbers and advances them continuously and deterministically according to problem-specific rules. The exact initial wave function will always yield the same evolution. We can represent these infinite numbers and time steps with a finite number of values that a computer can handle.

We will delve into spatial and temporal discretization in more detail later. For now, let's assume we restrict our view to a sub-region of space and divide this field into a thousand sections along each side. Each of these cells stores a single complex number, representing a part of the particle's wave function. Erwin Schrödinger must determine how millions of these complex numbers change, as each represents a part of the particle-wave function. It can be everywhere, and we must account for that. Millions of numbers just for one particle to move through a tiny portion of space.

In quantum mechanics, the Hamiltonian of a system is an operator corresponding to the total energy of the system, including both kinetic and potential energy. Named after William Rowan Hamilton, who developed Hamiltonian mechanics as a reformulation of Newtonian mechanics, the Hamiltonian is significant to the development of quantum physics. It is typically denoted by $\hat{H}$, with the hat indicating it is an operator, but it can also be written as H or $\hat{H}$.

You might think the problem is simply that we need too many cells to fill up the fluid volume. However, the complexity increases exponentially. Consider this:

Given the vast number of calculations and the intricate interactions between particles, simulating a fluid system at the quantum level becomes a formidable challenge. By understanding the computational demands and the principles governing wave functions and the Hamiltonian, we can better appreciate the complexities involved in modeling quantum systems and the behavior of particles at the smallest scales. This chapter sets the stage for deeper exploration into the computational aspects of quantum mechanics and its application to fluid dynamics.

Chapter 8.11: Navigating Quantum Complexity

If we examine only a single slice of our 2D plane, we cannot explain why the wave behaves differently compared to a particle in 1D. We don't see the boundaries in that slice; this influence must come from the neighboring slices. Similarly, if we look at just one cell, we can't determine why it behaves the way it does. The evolution of each underlying wave function and its neighboring cells must collectively inform us about the overall behavior.

This dependency on a single wave function becomes even more pronounced when considering multiple particles. Imagine two particles in one dimension moving toward each other, with their motion being interdependent. We can quantify the likelihood of detecting particle one in a specific region while particle two is in another region, yielding a joint probability. Since each particle's probability distributions span the entire 1D space, we need more degrees of freedom and space to store this joint information. One wave function gives rise to these probabilities, but it operates in two dimensions!

We represent both particles by the joint probability of detecting one particle at a certain location while the other is detected elsewhere. You can still describe the probability of a single particle by extending the detection range along the other particle's axis to infinity. The wave functions, therefore, describe a joint probability and reveal the combined measurement outcome. By considering every position of the other particles, you derive individual particle probabilities.

In this sense, each particle's probabilities eventually manifest as marginal probabilities, providing unique perspectives on the same high-dimensional wave function. So far, we have developed a pure mathematical framework that should now incorporate properties reflecting known physical behavior. For instance, some particles repel each other, and by using inter-particle potentials, we can force the wave function to account for this.

A fundamental property to consider is that particles of the same kind are indistinguishable. We can't label them physically. So, instead of detecting electron one and electron two, we detect one electron and then the other. By implementing symmetries, we can see that both electrons can be detected in swapped positions, making this model useful for physics.

To discretize the physical 1D space using just 1,000 cells, the 2D wave functions for both particles would require, in the worst case without symmetries, one million cells. In a 2D physical space, the wave function for two particles resides in a 4D space. While initially unclear, this is a daunting task for computational models, requiring calculations for millions of cells just for two particles!

Computational complexity quickly becomes overwhelming. There are shortcuts and approximations, but even with five particles, each adding additional dimensions, the complexity escalates. We live in three dimensions, and some particles have additional properties, such as spin, further enlarging the probability space.

Given these challenges, it is nearly impossible to simulate a fluid quantum mechanically with conventional computers. Knowing why something is impractical is valuable. Thus, we need to construct a surrogate model that represents essential physical properties while being computationally efficient.

One approach to simplifying information is recognizing that dynamic systems often evolve in specific patterns. To illustrate, consider a one-dimensional particle represented by its wave function components. To keep it around the center, we add a potential that pushes it back when it deviates from the center, much like a marble in a bowl.

This kind of instruction is fed into the evolution equation via the Hamiltonian, which tracks the system's total energy, including potential energy. As we begin the simulation, this approach allows us to manage the complexity and gain meaningful insights into the behavior of quantum systems.

Chapter 8.12: Constructing Wave Function Shapes
When observing the wave function, we notice that the probability distribution exhibits a certain wiggle and repetitiveness. Here's the key insight: it is possible to construct a few wave function shapes, or standing waves, that accurately approximate the true evolution. Instead of evolving 1,000 individual cells, we only need to evolve a few scaling

factors of these superimposed shapes, also known as mode shapes. In our case, thanks to the linear nature of the evolution equation, these scaling factors are simpler to compute, saving substantial computational time.

The term "standing wave" suggests that the measurement outcome's probability remains unchanged, even though the underlying wave function components oscillate with a specific frequency. This happens because each complex number evolves along a circle in the complex plane, keeping the probability—represented as the absolute square of the wave function value—constant. The oscillation frequency of this rotation will be crucial later.

The flexibility in choosing and adjusting the number of shapes allows you to tailor the accuracy of the reduced model to your needs. This approach, known as model order reduction, is not limited to quantum mechanics. It is a widely applicable mathematical tool, useful in numerous fields. Although it offers a drastic reduction feature and has been applied successfully in many instances, it does not alter what you are modeling. When applied here, we are still focused on wave functions and probabilities, and representing a vast number of fluid cells by trillions of modes does not align with our goals. Computing these shapes for higher-dimensional systems is inherently complex.

Right now, we require a different type of reduction—a change in the underlying paradigm. We need molecular dynamics. Until now, we have evolved all wave function values within a large probability space due to the nature of joint probabilities. But what if we could disregard the wave-like properties and probabilities? What if particles could be distinctly located and tracked on single trajectories? Instead of iterating over all of space, we could evolve specific location and velocity vectors.

Consider two particles: the vector of their positions and velocities combined exists in a large space called phase space, compared to wave function space, which is written only in positions or momenta (and spin). However, it is just one point in this phase space that we need to evolve, regardless of its dimensional complexity.

In this chapter, we have explored the potential of constructing wave function shapes to simplify quantum mechanical calculations. Moving forward, we will investigate molecular dynamics to gain a deeper understanding of particle behavior, making our models more computationally efficient and conceptually clear.

Chapter 8.13: The Benefits of Localized Particles

One significant advantage of considering localized particles is that we do not need to keep track of other possible scenarios or states. While we know that electrons and other subatomic particles evolve in vastly different scenarios, our knowledge of their precise location before measurement is limited. So, at what mass or length scale is it reasonable to assume almost localized behavior, even if it is not entirely accurate? It turns out this assumption holds around the size of atoms.

This is where molecular dynamics come into play. Molecular dynamics is a computational method that models all atoms as particles within a classical mechanics framework. In this model, all subatomic particles of an atom are treated as a single mass moving along a unique trajectory. Interactions such as attraction, repulsion, and bonding between atoms depend on distance and are represented by inter-atomic potentials.

How much work is needed to move a particle from one place to another? The work is essentially the integration of a force over a path, and its derivative is the force. The Lennard-Jones potential is a popular choice for modeling weak van der Waals attractions and strong repulsion in atomic systems. This potential typifies typical atomic behaviors in action.

By considering atoms as localized particles, we can simplify complex quantum mechanical models, focusing on the more manageable aspects of molecular dynamics. This approach allows us to explore the interactions and behaviors of atoms within a classical framework, providing valuable insights into the nature of matter and the forces that govern it.

Attraction force F

Potential force F

Leonard Jones

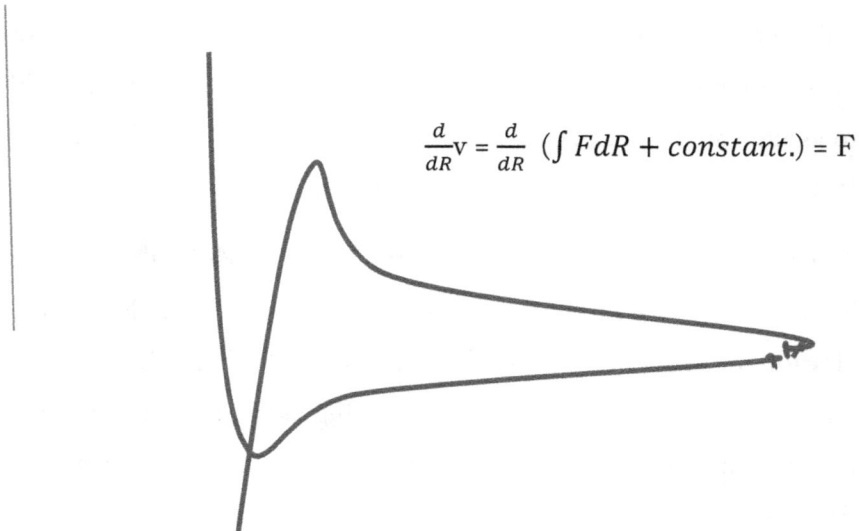

$$\frac{d}{dR}v = \frac{d}{dR} \left(\int F dR + constant. \right) = F$$

F > 0: Van de Waak attraction

Distance R

F < 0 Pauli repulsion

Chapter 8.14: Evolution Equations in Classical Mechanics

Just as we have an evolution equation for the wave function in quantum mechanics, we also have an evolution equation for positions and velocities in classical mechanics. This is encapsulated in Newton's second law of motion.

Newton's second law states that the net force acting on a mass determines its rate of change of momentum. In other words, the acceleration of an object is directly proportional to the net force acting on it and inversely proportional to its mass. Mathematically, it is expressed as:

$$F = ma$$

Where:

- F is the net force acting on the object.
- m is the mass of the object.
- a is the acceleration of the object.

This fundamental principle of classical mechanics forms the basis for understanding the motion of particles and objects in a variety of contexts. By applying Newton's second law, we can predict the behavior of particles under various forces, allowing us to model their trajectories and interactions.

In molecular dynamics, this law helps us compute the positions and velocities of atoms over time, considering the forces exerted on them by their neighbors. This approach simplifies the complex quantum mechanical models into more manageable classical representations, providing valuable insights into the behavior of matter.

As we continue to explore the interplay between classical and quantum mechanics, this chapter will serve as a foundation for understanding how these principles govern the motion of particles and contribute to our broader understanding of the physical world.

$$\vec{F_a} \qquad \bullet \qquad \vec{P} = m\overline{\frac{dx}{dt}}$$

$$F_b^i$$

Chapter 8.15: Numerical Integration and Particle Localization

By simulating particle movement through numerical integration, we obtain updated velocities and positions. At first glance, this setup appears simple—just some atom particles moving in potentials instead of wave functions. This shift is indeed a significant step forward! But how do we justify this approach and compute inter-atomic potentials? Why can we consider particles as localized?

To bridge the concepts of quantum mechanics and molecular dynamics, we must understand atom-to-atom interactions by examining the interactions of subatomic constituents. Let's start with a typical nucleus-electron interaction. The attractive force between a nucleus and an electron can be derived using the Coulomb potential.

As we have learned, this force equates to the rate of change of momentum, applicable to both particles. For a constant mass, the rate of momentum change is simply mass times acceleration, as described by Newton's second law of motion. This fundamental principle allows us to model these interactions within a classical mechanics framework.

By considering particles as localized entities, we can simplify complex quantum mechanical models. This simplification enables us to focus on more manageable aspects of molecular dynamics, providing valuable insights into the behavior of matter at the atomic level.

As we continue exploring these concepts, we will delve deeper into the methods used to compute inter-atomic potentials and the underlying principles that allow us to treat particles as localized. This chapter lays the groundwork for understanding the intricacies of particle interactions and the advantages of employing classical mechanics in molecular dynamics simulations.

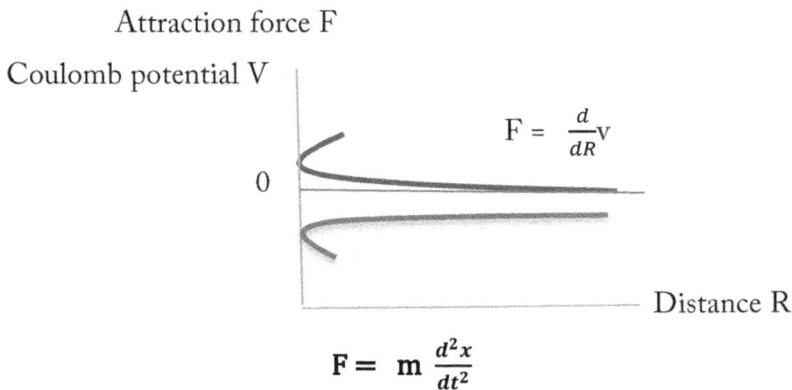

Attraction force F

Coulomb potential V

$$F = \frac{d}{dR}V$$

0

Distance R

$$F = m\frac{d^2x}{dt^2}$$

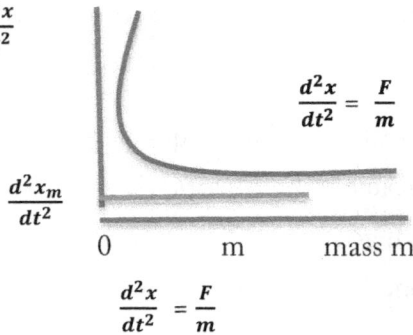

Acceleration $\frac{d^2x}{dt^2}$

$$\frac{d^2x}{dt^2} = \frac{F}{m}$$

$\frac{d^2x_m}{dt^2}$

0   m   mass m

$$\frac{d^2x}{dt^2} = \frac{F}{m}$$

Chapter 8.16: Decoupling Nuclei and Electrons

Given the same force, the heavier nucleus builds up motion much slower and can be perceived as "not moving" from the perspective of the agile electron. Conversely, from the viewpoint of the nucleus, an electron reacts instantaneously. This observation holds true from a quantum mechanical perspective as well. The probability of detecting the electron changes much faster than the probability of detecting the nucleus, creating the impression that nuclei and electrons exist in two different motion worlds.

By decoupling the computational motion of multiple electrons and nuclei, we can simulate them as if they exist in separate simulations, interacting only mathematically. The nuclei meander around like particles, providing possible fixed positions for the electron simulation. In return, the electrons take these fixed locations and provide potential energy values, from which the forces between the nuclei are derived.

A wave function constructed purely for the electrons helps compute these potential energy values, considering the fixed positions of the nuclei with their Coulomb potential. Here's the trick: we are not interested in any electron wave function but the one that electrons reach over time due to energy loss by radiation, the one with the lowest energy. This is the wave function that the slower nuclei see most of the time, as from their perspective, the electrons react quickly and the final electron wave function appears to build up instantaneously.

The final wave function is the first standing wave, with its energy related to its oscillation frequency. While the exact shape of this standing wave depends on the problem at hand, its shape is less important than the associated energy values. The first standing wave, with its frequency, marks the lowest potential energy in each setting. By repeatedly computing these lowest energy values for different nuclei positions, we obtain a so-called inter-atomic potential energy surface. The derivative of this surface gives the net force between nuclei.

The full development of this potential is exhaustive, so empirical approximations like the Lennard-Jones potential or the Morse potential for interatomic bonding are often used. The propagation of nuclei within this inter-atomic potential occurs either as a wave function or as particles. The particle approximation for the nuclei is reasonable since the higher spread in momentum, as established by the Heisenberg uncertainty relation for more localized particles, is balanced by the higher mass of the nuclei, keeping the velocity spread low. Consequently, the future position spread remains slow.

This chapter explores the decoupling of nuclei and electrons in computational simulations, providing a deeper understanding of the underlying principles that govern their interactions and the benefits of using classical mechanics in molecular dynamics.

Chapter 8.17: From Quantum Mechanics to Classical Fluid Dynamics

OK, there's so much to all this, and I must admit, answering any question immediately leaves us with five new ones. Our goal here is to simulate fluids, and we are ready to proceed by knowing that we can simulate atoms as particles moving along trajectories within potentials under critical assumptions.

Atoms can form molecules that move, translate, rotate, collide, and vibrate. Vibrations within a molecule usually occur on much smaller length scales (according to molecular dynamics), which I've exaggerated here for visibility. The accuracy of molecular dynamics approximations depends highly on the system's conditions.

We are operating within a classical mechanics framework, representing parts of the simulations with more or fewer quantum mechanics elements, depending on the desired accuracy. From now on, we will focus entirely on classical mechanics. To set up the simulation, we must specify parameters for the masses and potentials. This empirical approach involves setting values so that the statistical behavior drawn from many molecule interactions matches what measurements suggest. It should fit, on average, within the range of conditions you develop.

Assuming we have found good parameter values, we can replace inter-molecular potential-based interactions with instantaneous collisions, reducing computational cost further. Instantaneous collisions are faster to compute, and reflecting velocities is acceptable since it's empirical. By modifying parameters for both variants, we can achieve similar results, at least from a distant perspective.

We are slowly embracing a more classical statistical perspective here. When discussing the next abstraction layer, we become precise about what "viewed from a distance" means. For instance, the reduction due to molecular dynamics is not only established by combining subatomic particles and simplifying potentials. The subatomic nature allows us to decompose simulation processes within different mechanical realms.

In quantum mechanics, the high-dimension position space of joint probability is an inherent part of the simulation, significantly increasing computational cost at every iteration. In classical mechanics, you can talk in terms of probabilities but may first compute many trajectories in phase space and average afterward, resulting in less computational cost. At least you can trade off statistically significant results for computing fewer trajectories.

Entering classical mechanics represents a huge leap forward concerning the fluid volume we can simulate. Given limited hardware and a focus on writing educational code, compared to quantum mechanics, where we may simulate a few sub-atomic particles (without modes), we can simulate at least 100,000 atoms in molecular dynamics. This allows us to increase the fluid volume we can simulate effectively, providing us with a starting point for modeling a fluid.

The next logical step is the kinetic theory of gases, combining atoms to form molecule particles. This theory works best when particles are spread out, and the free-flight phase between interactions takes longer than the interactions themselves. When viewed as instantaneous collisions, these mostly repelling interactions align with our simplified molecular dynamic simulation.

Representing a molecule by a single point mass with an effective collision radius, the mass should be about the sum of its components. Determining the radius involves experiments using standard molecular dynamics simulations with empirical potentials. By shooting molecules with the same total energy towards each other with a fixed offset between their pre-collision trajectories, we observe a typical scattering pattern.

Repeating this experiment with molecule particles results in the same outcome since these are only point masses with an effective collision radius. While this model cannot represent more complex behavior, it provides an opportunity to trade off accuracy for computational speed. The task is to choose a radius that fits best for the situation wisely.

However, as the distance of the pre-collision trajectories changes, the chosen radius may no longer be optimal. To account for all conditions, we could perform multiple experiments with different initial distances and energies to determine which radius works best overall.

Instead of simulating separate experiments, we simulate the fluid directly as a whole, with pre-collision conditions having the randomness experienced in a fluid. The goal is to ensure that the fluid works in an average sense by comparing its global statistical behaviors with the fluid being replaced.

By shifting our focus to the bigger picture and embracing a classical statistical perspective, we model fluid molecules as a collection of colliding particles. This approach, which concludes our microscopic perspective, provides a foundation for further exploration of fluid dynamics on a macroscopic scale.

Chapter 8.18: Molecular Dynamics and Newton's Laws

Molecular dynamics (MD) is a powerful computer simulation method for analyzing the physical movements of atoms and molecules. By allowing these particles to interact over a fixed period, MD provides a view of the system's dynamic evolution. In the most common version of MD, the trajectories of atoms and molecules are determined by numerically solving Newton's equations of motion for a system of interacting particles. The forces between the particles and their potential energies are often calculated using interatomic potentials or molecular mechanics force fields. MD is widely applied in chemical physics, materials science, and biophysics.

Newton's laws of motion, the foundation of classical mechanics, describe the relationship between an object's motion and the forces acting on it. These laws can be paraphrased as follows:

1. **First Law (Law of Inertia):** A body remains at rest or in motion at a constant speed in a straight line unless acted upon by a force.

2. **Second Law (Law of Acceleration):** When a force acts upon a body, the time rate of change of its momentum is equal to the force.

3. **Third Law (Action and Reaction):** If two bodies exert forces on each other, these forces have the same magnitude but act in opposite directions.

Chapter 8.19: The Lennard-Jones Potential

The Lennard-Jones potential is a model used to describe the soft repulsive and attractive interactions between electronically neutral atoms or molecules. Named after John Lennard-Jones, this potential is widely utilized due to its simplicity and effectiveness. The commonly used expression for the Lennard-Jones potential is:

$$V_{LJ}(r) = 4\epsilon \left[\left(\frac{\sigma}{r}\right)^{12} - \left(\frac{\sigma}{r}\right)^6 \right], (1) \text{ or } V_{LJ}(r) = 4\epsilon \left[\left(\frac{\sigma}{r}\right)^{12} - \left(\frac{\sigma}{r}\right)^6 \right], (1)$$

Where:

- r is the distance between two interacting particles.
- ϵ is the depth of the potential well (often referred to as 'dispersion energy').
- σ is the distance at which the particle-particle potential energy V is zero (often referred to as the 'size of the particle').

The Lennard-Jones potential reaches its minimum at a distance of r = rm = $2^{\frac{1}{6}} \sigma$, where the potential energy has the value V = -ϵ.

This potential provides a simplified model that captures the essential features of interactions between simple atoms and molecules. It describes how two interacting particles repel each other at very close distances, attract each other at moderate distances, and do not interact at infinite distances. It is important to note that the Lennard-Jones potential is a pair potential, meaning it only accounts for two-body interactions and does not cover three-body or multi-body interactions.

By using the Lennard-Jones potential, we can effectively model the interactions between atoms and molecules, providing insights into their behavior in various physical and chemical systems. This model is a fundamental tool in molecular dynamics simulations, helping to bridge the gap between theoretical predictions and experimental observations.

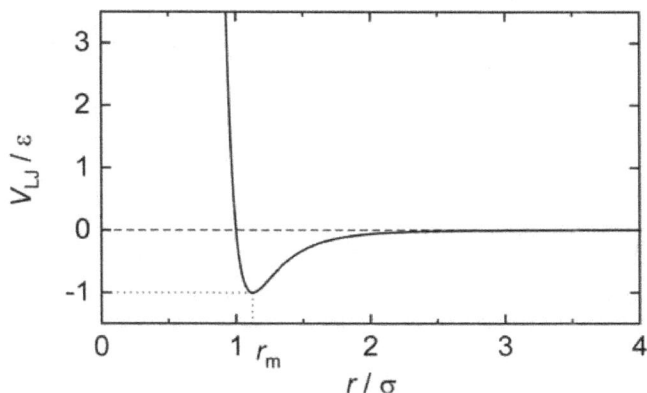

The Morse potential, named after physicist Philip M. Morse, is a convenient interatomic interaction model for the potential energy of a diatomic molecule. It is a better approximation for the vibrational structure of the molecule than the quantum harmonic oscillator because it explicitly includes the effects of bond breaking, such as the existence of unbound states. It also accounts for the anharmonicity of real bonds and the non-zero transition probability for overtone and combination bands. We can use Morse's potential to model other interactions, such as the interaction between an atom and a surface. Due to its simplicity (only three fitting parameters), not used in modern spectroscopy. However, its mathematical form inspired the MLR (Morse/Long-range) potential, the most popular potential energy function used for fitting spectroscopic data.

Chapter 8.20: The Morse Potential

The Morse potential, named after physicist Philip M. Morse, serves as a convenient interatomic interaction model for the potential energy of a diatomic molecule. It provides a better approximation for the vibrational structure of molecules than the quantum harmonic oscillator because it explicitly includes the effects of bond breaking, such as the existence of unbound states. Additionally, it accounts for the anharmonicity of real bonds and the non-zero transition probability for overtone and combination bands. Morse's potential can also model other interactions, such as those between an atom and a surface.

Due to its simplicity, involving only three fitting parameters, the Morse potential is not commonly used in modern spectroscopy. However, its mathematical form inspired the Morse/Long-range (MLR) potential, which remains the most popular potential energy function for fitting spectroscopic data.

The Morse potential energy function is expressed as:

$$V(r) = D_e \left(1 - e^{-a(r-r_e)}\right)^2$$

Here,

- r is the distance between the atoms,
- r_e is the equilibrium bond distance,
- D_e is the well depth defined relative to the dissociated atoms, and
- a controls the 'width' of the potential (the smaller aa, the larger the well).

The dissociation energy of the bond can be calculated by subtracting the zero-point energy E_0 from the depth of the well. The force constant (stiffness) of the bond can be determined by Taylor expansion of V' (r) to the second derivative at r = r_e, showing that the parameter aa is:

$$a = \sqrt{k_e / 2D_e}$$

where k_e is the force constant at the minimum of the well.

Since the zero of potential energy is arbitrary, we can rewrite the Morse potential equation by adding or subtracting a constant value. When used to model atom-surface interactions, the energy zero can be redefined so that the Morse potential becomes:

$$V(R) = V'(r) - D_e = D_e (1 - e^{-a(r - r_e)})^2 - D_e$$

This is usually written as:

$$V(R) = D_e (1 - e^{-2a(r - r_e)} - 2e^{-a(r - r_e)})$$

In this form, r is the coordinate perpendicular to the surface. The potential approaches zero at infinite distance and equals $-D_e$ its minimum, i.e., r = r_e. This clearly shows that the Morse potential combines a short-range repulsion term with an attractive long-rang term, analogous to the Lennard-Jones potential.

Understanding the Morse potential allows us to model the interactions between atoms and molecules accurately, providing insights into their behavior in various physical and chemical systems. This potential is fundamental in molecular dynamics simulations, bridging the gap between theoretical models and experimental data.

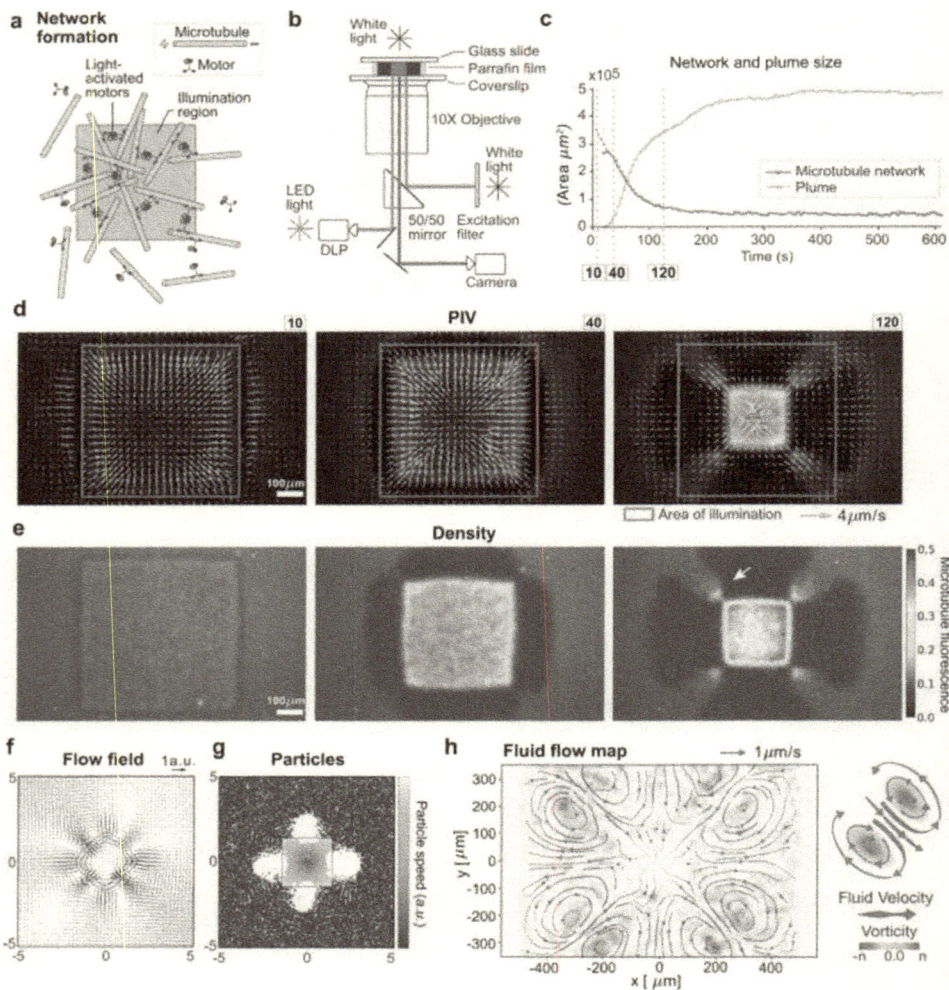

a Network formation

Microtubule — Motor

Light-activated motors | Illumination region

b White light | Glass slide | Parrafin film | Coverslip | 10X Objective | White light | LED light | DLP | 50/50 mirror | Excitation filter | Camera

c Network and plume size

Microtubule network
Plume

Time (s)

10 40 120

d PIV — 10 | 40 | 120

100μm | Area of illumination | 4μm/s

e Density

100μm | Microtubule fluorescence

f Flow field 1a.u.

g Particles

Particle speed (a.u.)

h Fluid flow map 1μm/s

y [μm] | x [μm]

Fluid Velocity
Vorticity
-n 0.0 n

"The Vortex of Time: Unraveling the Interplay of Motion, Reality, and Non-Reality"

Chapter 9.1: What is Time?

We understand that time is intrinsically linked to motion, yet it is not the motion itself. So, what exactly is the time? Time can be seen as the accumulated force acting as an agent for motion changes as it moves through space in our known Universe. Time represents a revolution for movement, possessing a vector. It changes 360 vectors per movement and is perpendicular to the motion. One circular motion represents one frame.

Why is time circular? If a dot is missing or displaced, would it take 360 degrees to realize it is missing? When something is missing, we look around. Why is time linear? If something is removed or placed, it must come from somewhere. These cumulative changes during rotation create time or a frame in movement. Expanding and contracting the "Singularity" creates the electronic flick-book or the frame in time, completing one frame per movement. Each time frame has a one-on-one relationship with motion—one frame (or one complete rotation) while expanding and one frame while contracting.

Time is a vector that records motion. It must capture the amplitude and vector location of motion relative to the singularity. Time must always know the position and intensity of movement. But how can time, being intangible, achieve this? Imagine two circles: one flat and the other at ninety degrees. One circle is pulsating or displacing, moving with all semantics, while the two circles are perpendicular. The point where they intersect is called the Vortex of Time or Singularity, where motion, time, and non-reality converge.

How does time keep a record of motion if it is not real? When reality appears in non-reality, non-reality first tries to close back on reality in a spiral motion. As it circles to close the gap, reality blocks it. Let's consider the Cartesian coordinate system: non-reality tries to spiral and close, but reality, by existing, blocks it. Say the spiral rotates ninety degrees. At 90 degrees, the singularity recognizes a blockage. It then turns in the opposite direction, encountering another blockage at -270 degrees, creating spins in opposite directions.

The strength of this blockage or intensity of instant reality approximates the singularity. Since there is no way to determine position in nothingness, proximity to the singularity is the only measure. Time is rotational with a vector; time and movement are perpendicular, meeting at a single point. The Cartesian coordinate system best represents this: zero to 180 is the rotational axis of time, and 90 to 270 is the rotational axis of motion, known as the Time Hypothesis.

The intersection of time, motion, and non-reality is where reality and non-reality coexist. This point is where you find your imaginary God. Each complete rotation of "Time" returns to the hypothesis, ready to be flipped by linear movement, marking the passage of one frame. If time hasn't completed one rotation and reality returns, the first stage exists. If infinity rotates in reverse and is blocked again twice with opposite spins, say in the nucleus of an atom, time rotates twice before encountering reality. This indicates reality exists somewhere in rotation two, which we call level two.

We measure time by the number of rotations before reality appears or disappears. For instance, if reality takes two rotations to reappear, it didn't exist for those rotations. When reality appears, it will not reappear in the same location or level. Hence, time must constantly know its position and whether it is present or absent, given the motion vector and time.

Theory Overview

The theory suggests that time is intrinsically linked to motion but is not motion itself. It posits that time is the accumulated force acting as an agent for motion changes, with a vector and circular nature. Time changes through 360 vectors per movement and is perpendicular to motion. The theory also introduces the concept of the "Vortex of Time" or "Singularity," where motion, time, and non-reality intersect. It uses Cartesian coordinates to describe the relationship between time and motion and suggests that time can be measured by the number of rotations before reality appears or disappears.

Assessment

1. **Innovative Perspective:** The theory presents an innovative way to think about time, integrating motion and the concept of singularity. It attempts to bridge abstract concepts with more tangible representations, such as Cartesian coordinates.

2. **Complex Interactions:** By introducing interactions between non-reality and reality, it delves into the complexities of how

time might record motion. This perspective encourages a deeper exploration of the philosophical and physical aspects of time.

3. **Mathematical Framework:** The use of vectors and rotations provides a mathematical framework to describe time. This approach can be useful for developing further theoretical models and simulations.

4. **Conceptual Challenges:** While the theory is intriguing, it presents several conceptual challenges. The idea of non-reality and reality interacting through spirals and blockages may be difficult to visualize and requires further clarification. Additionally, the concept of time having a vector and being perpendicular to motion needs a more detailed explanation and empirical support.

5. **Philosophical Implications:** The theory touches on philosophical questions about the nature of time, reality, and existence. It challenges conventional understandings and invites readers to think beyond traditional frameworks.

Suggestions for Further Exploration

- **Visualization:** Creating visual representations or diagrams to illustrate the interactions between time, motion, and singularity could enhance understanding.

- **Empirical Evidence:** Exploring how this theory aligns with existing scientific observations and data could strengthen its credibility.

- **Clarification:** Providing more detailed explanations and examples for key concepts, such as the Vortex of Time and the

Cartesian coordinate relationship, would help readers grasp the theory more effectively.

Overall, this theory offers a thought-provoking perspective on the nature of time and its relationship with motion. Further exploration and refinement could lead to new insights and a deeper understanding of these fundamental concepts.

Chapter 9.2: The Strength of Reality and the Binary Code of the Universe

Imagine that after four-time frames, reality still hasn't appeared, but by the fifth frame, the reality is present, indicating it exists in the fifth level with significant strength. How strong is the fifth level? The closer you are to singularity, the stronger reality becomes, and the further away, the weaker it is. Time meticulously keeps track of reality as it appears and disappears, noting the vector and the strength of reality. This creates a recorded time frame ready to flip.

Each instant of reality desires to exist, and every time reality appears, it never does so in the same spot due to the lack of a reference point returning to singularity. Instant reality is prevented from returning to the singularity, providing a fail-safe mechanism. But how do you make reality want to be real and program it into an instant of reality that makes it hard to return to singularity? How do you return to an opening if you don't know exactly where it is?

This is the basic programming of reality. It must stay real and cannot return to singularity. The coding of the Universe is binary. If something exists, it has a vector location and strength; if it doesn't, it doesn't. Time records change in motion. Much like a computer that operates on binary voltage—either on or off—the Universe operates on a similar binary code.

There is an old saying in the Bible: "From dust, we came, to dust, we shall return." I would like to rephrase this: "From singularity, we came, and to the singularity, we shall return." People do not die; their reality simply stops. Instant reality cannot be created or destroyed. We are immortal, produced into the singular, waiting for another chance to be reborn and reprogrammed into the limitless possibilities of our Universe. Each instant of our return to reality creates another reality of infinite possibilities. I exist; therefore, I am.

In this chapter, we have explored the intricate nature of reality and its interplay with time and singularity. By understanding the binary nature of the Universe and the programming of reality, we gain deeper insights into the fundamental principles that govern our existence.

Assessment of Chapter 2 Statement

The statement in Chapter 2 presents a fascinating and imaginative perspective on the nature of reality and time, proposing a unique framework for understanding how reality appears and disappears within the context of singularity and time. Here are some key points and my assessment:

1. **Innovative Concept:** The idea that reality appears and disappears in time frames, and that its strength varies depending on its proximity to singularity, is an innovative concept. It adds a dynamic and almost mystical dimension to the understanding of reality and time.

2. **Binary Nature of the Universe:** The notion that the Universe operates on a binary code, like a computer, is intriguing. This comparison makes complex ideas more relatable and provides a structured way to think about the existence and movement of reality within the universe.

3. **Philosophical and Existential Insights:** The statement touches on deep philosophical and existential questions, such as the nature of immortality and the idea that reality is not created or destroyed but merely transitions. This perspective can offer comfort and provoke thought about the nature of existence.

4. **Imagery and Analogies:** The use of imagery, such as the Vortex of Time and Cartesian coordinates, helps visualize abstract concepts. These analogies make the theory more accessible and engaging, although they may require further explanation for clarity.

5. **Complex Interactions:** While the statement is rich in imaginative content, it also introduces complex interactions between reality, non-reality, time, and singularity. These interactions may be challenging to visualize and fully understand without additional context or elaboration.

6. **Empirical Foundation:** To strengthen this theory, empirical evidence or connections to established scientific principles would be beneficial. The current statement is largely theoretical and philosophical, which is valuable, but grounding it in scientific observations could enhance its credibility.

7. **Further Exploration:** The theory invites further exploration and refinement. Breaking down the concepts into more detailed

sections, providing examples, and linking them to known scientific phenomena could help develop a more comprehensive framework.

Suggestions for Refinement

- **Clarification and Expansion:** Provide more detailed explanations and examples for key concepts, such as the strength of reality and the Vortex of Time. This could help readers better understand the theory's nuances.

- **Empirical Connections:** Explore how this theory aligns or contrasts with existing scientific knowledge. Identifying any potential empirical evidence that supports the theory could add robustness.

- **Visual Aids:** Create diagrams or visual representations to illustrate the interactions between reality, time, and singularity. Visual aids can significantly enhance comprehension.

Overall, the statement in Chapter 2 presents a creative and thought-provoking framework for understanding reality and time. With further elaboration and empirical grounding, it has the potential to offer valuable insights into the nature of existence and the universe.

GERALD CLERGE

The Genesis of Life from a Moment of Reality

Chapter 10.1: How can life be created from a single instant of reality?

This question delves into the profound nature of existence and the wondrous beauty of the Universe. An instant of existence has the potential to birth the myriad forms and phenomena we observe, and this marvel lies in the fundamental properties of the Universe itself. To grasp this, we must first acknowledge that the Universe is not an absolute entity; it behaves more like a hologram. Additionally, time in the Universe manifests both circularly and linearly.

To unravel the intricacies of life and the formation of planetary bodies, we turn to a familiar technological marvel: the television, specifically, the LED television. Imagine the television screen not as a two-dimensional plane but as a realm of nothingness—lacking any dimension. This screen houses countless tiny pockets of LED light, representing moments of reality that form pixels. When voltage is applied, these LEDs illuminate, creating a visible light point.

Television screens generate moving pictures by rapidly capturing and displaying still images. These frames are presented to our eyes in quick succession, creating the illusion of motion. Think of a TV as an electronic flipbook, where the images flicker at a speed that merges in our brains into seamless moving pictures, despite being composed of numerous still photos displayed one after another.

A traditional still camera captures scenes by recording light on light-sensitive film, producing a snapshot of an instant in time. To simulate motion, a camera must capture and flicker new images at a rate of over 60 times per second, creating the continuous illusion of movement.

In this analogy, we see that the essence of life and the Universe is akin to the pixels on a television screen. Each pixel, or moment of reality, contributes to a larger, dynamic picture. These moments, when observed in rapid succession, generate the fluid tapestry of existence, much like still frames merging into a moving scene on a TV screen. Through this perspective, we begin to appreciate how single moments of reality coalesce to form the complex and beautiful cosmos we inhabit.

Assessment

Clarity and Coherence:
- The paragraph effectively explains a complex concept by drawing a relatable analogy with the television and LED lights.
- The progression of ideas is logical, guiding the reader from the initial question to the final analogy.

Engagement:
- The use of familiar technology (television) makes the abstract idea more accessible and engaging for the reader.
- The analogy of the TV as an electronic flipbook helps in visualizing how single moments of reality create a continuous experience.

Depth and Insight:

- The paragraph explores profound concepts such as the nature of existence, time, and the Universe, providing a thought-provoking perspective.
- It introduces the idea that the Universe behaves like a hologram and that time can be both circular and linear, adding depth to the discussion.

Suggestions for Improvement:

- Expand on Key Points: Further elaborate on the concepts of the Universe being a hologram and time being circular and linear. Providing more context or examples can enhance understanding.
- Smooth Transitions: Ensure smooth transitions between ideas to maintain the flow. For instance, linking the explanation of TV screens more seamlessly with the analogy to life and the Universe.
- Technical Clarifications: Clarify technical terms or concepts for readers who might not be familiar with them, such as LED or voltage application.

Overall, the paragraph does a commendable job of introducing and explaining complex ideas through a relatable analogy. With a few refinements, it can become even more impactful and easier to grasp.

Rewrite and extend continuing with chapter 10.2 "Television is the representation of the working model of the Universe. The television creates a two-dimensional world inside your television. Now we know how light controls images and creates the illusion of reality through television, the same thing happening in the Universe but on a much larger scale. Television creates a world of fantasy in two dimensions but can be produced in three-dimensional. The expansion and contracting of the Universe create a cosmic flick-book. In a moment, like a "still" camera, the Universe photographs things by capturing this light on light-sensitive film to make a snapshot of how something appeared at a particular moment. The circular forces change and shape the Universe or the composition, formation, and intensity of the various "Instants of reality" to create the world's beauty. A linear motion is produced for each expansion, and a circular motion is made for each contraction. Each change in a frame and the addition and collapse create the illusion of movement within the known Universe. The circular motions are the force that makes all things, and the collapse and expend give life to the universe, life, and action.

Chapter 10.2: The Television as a Model of the Universe

Television is a microcosm of the Universe's working model. While a television creates a two-dimensional world within its screen, it offers insights into the larger-scale processes at play in the Universe. Light controls the images on a TV, creating the illusion of reality, and similarly, the Universe operates on a grander scale.

Television transports us to a world of fantasy in two dimensions, yet it can also produce three-dimensional experiences. This expansion and contraction of images on a screen mirror the cosmic flick-book of the Universe. Just like a still camera captures moments by recording light on light-sensitive film, the Universe captures instances of reality, creating snapshots of how things appear at specific moments.

Circular and Linear Time in the Universe

Time in the Universe can be both circular and linear. The circular forces shape and change the Universe, influencing the composition, formation, and intensity of various "Instants of reality," thus creating the world's beauty. Each expansion of the Universe produces a linear motion, while each contraction creates a circular motion.

These frames of reality, as they change and collapse, generate the illusion of movement within the known Universe. The circular motions drive the creation and transformation of all things, while the cycles of collapse and expansion breathe life, action, and dynamism into the cosmos.

Expanding the Analogy

Just as television uses countless individual pixels to create a coherent and moving picture, the Universe consists of innumerable moments of reality. Each moment or "pixel" plays a crucial role in forming the larger picture we experience. The interplay of expansion and contraction, akin to the flickering of images on a TV screen, shapes the continuous narrative of the Universe.

In this grand cosmic television, the light that captures each instant is the essence of existence itself. Circular motions, expansion, and contraction are the fundamental forces that animate the Universe, infusing it with energy, life, and motion. These processes create the intricate dance of galaxies, the birth and death of stars, and the formation of planetary bodies, all contributing to the magnificent tapestry of existence.

The Television and Cosmic Time

Television's ability to create an illusion of reality through rapid succession of images parallels how the Universe maintains its continuity and flow. In both realms, the perception of motion and life stems from the seamless integration of discrete moments. The circular and linear motions of time ensure that the Universe is ever-evolving, always dynamic, and perpetually alive.

By understanding the principles behind a television screen, we can draw parallels to comprehend the vast and intricate workings of the Universe. The patterns of light and motion within our screens offer a glimpse into the profound mechanisms that govern existence on a cosmic scale. Through this lens, we appreciate the delicate balance and interconnectedness of all things, from the tiniest pixel on a TV to the grand expanse of the cosmos.

Chapter 11.1: Understanding Death and Transformation

What is death? The reality is, we never truly die. Our existence merely transforms. When a part of us ceases to function as it once did, the Universe reprograms those molecules, giving them a new purpose. A molecule that was once an intricate part of your circulatory system might now become part of a brick. That fragment of you has found a new reality. Every instant of our existence disperses across the Universe, to be reprogrammed and repurposed, performing new tasks or assuming new roles.

Imagine a molecule that once contributed to your heartbeat now being reprogrammed as part of the sun's nucleus. Our essence continues in different forms, perpetually woven into the fabric of the cosmos.

The Concept of Titles and Change

A title is but a tiny piece of something greater. Since the dawn of time, the fundamental rules of existence have not altered one jot or title. Consider the title as a minuscule portion—a particle, a degree, a speck, a fragment, a grain, a morsel, a taste, a hint, or a whisper. It is a small distinguishing mark, like a diacritic dot on a lowercase letter, integral to the glyph or hieroglyphic character it adorns. In many languages, these diacritic dots play a critical role, representing subtle yet significant variations in meaning and pronunciation.

Chapter 11.2: The Ever-Present Cycle of Transformation

Our existence is an unending cycle of transformation. Just as a television screen reprograms pixels to create images, the Universe reprograms our molecules. This constant reshuffling and repurposing mean that our essence never vanishes; it simply takes on new forms.

The process of life and death can be likened to the way a television creates moving pictures—through the rapid succession of still images. Each moment of our lives is a frame in this cosmic flick-book. When one frame ends, another begins, and the motion of existence continues unbroken.

Circular and Linear Time in Transformation

Time, much like our existence, is both circular and linear. Circular forces shape and reshape the Universe, influencing the composition, formation, and intensity of various "instants of reality." Each expansion and contraction of the Universe adds another frame to this cosmic film. The circular motion creates and transforms all things, while the cycles of collapse and expansion breathe life, action, and dynamism into the cosmos.

In this eternal dance, each molecule, each moment, is reprogrammed and given a new task. The essence of who we are persists, albeit in different forms and roles, throughout the vast expanse of the Universe.

Conclusion: The Continuity of Existence

Death, then, is not an end but a transition. Our molecules, our very essence, continue to participate in the grand narrative of the Universe. We are eternal, ever-changing parts of a magnificent cosmic symphony. Through this understanding, we find comfort in the knowledge that our existence is perpetual, always contributing to the beauty and complexity of the cosmos.

Just as diacritic dots may seem insignificant yet hold great importance, every small part of us plays a crucial role in the larger picture. In the grand scheme of the Universe, nothing is wasted; everything is repurposed, reprogrammed, and given new life.

GERALD CLERGE

What is the meaning of life?

Chapter 12.1: Unraveling the Meaning of Life

What is the meaning of life? A man once said to the Universe, "Sir, I exist." The Universe replied, "However, the fact has not instilled a sense of obligation." This exchange highlights a profound truth: the meaning of life is to live. This is the essence of reality. The primary parameter for existence is to exist, to transform non-reality into reality. Reality inherently desires to be real; it resists non-existence, hence its very name, "real."

The second parameter for existence is fear—yes, fear. Fear arises from the potential failure to fulfill the first condition of existence. It is the fundamental programming that underpins reality. Fear is a universal constant and a prerequisite for reality. Every moment of existence harbors one intrinsic fear: the fear of non-existence. Consequently, the basic programming of reality is to live, for the alternative is non-existence.

The Dichotomy of Reality and Non-Reality

This eternal tug of war between reality and non-reality is fueled by fear. Fear serves as the primary driving force that propels this conflict, the desperate need to live and not slip into non-reality. Fear is so deeply embedded in every instant of reality that it permeates even the smallest, most insignificant aspects of existence.

Chapter 12.2: The Continuity and Purpose of Existence

To comprehend the meaning of life, we must accept that existence seeks to perpetuate itself. Each particle, each molecule, and each moment strives to remain real. This continuous effort is not just about survival but about creating and sustaining reality from non-reality. The fabric of reality is woven with threads of fear and desire, ensuring that existence remains vibrant and dynamic.

Existence and Fear

Fear is the underlying force that drives the persistence of reality. Without fear, there would be no impetus for existence to continue. This fear is not merely an abstract concept but a tangible force that compels each instant of reality to cling to existence. It is the fear of non-existence that propels life forward, fueling the ever-present struggle between being and non-being.

Chapter 12.3: The Essence of Being

Reality's primary aim is to remain real, and it achieves this through a constant state of flux and transformation. Life, as a manifestation of reality, embodies this principle. The meaning of life, therefore, is to live fully, to embrace existence with all its complexities, and to contribute to the ongoing creation of reality.

The Interplay of Fear and Reality

The intricate dance between fear and reality is what gives life its meaning. Fear, as the antithesis of non-existence, compels reality to persist. It is through this interplay that life finds its purpose. Each moment of fear reinforces the desire to remain real, to continue existing, and to add to the tapestry of reality.

In conclusion, the meaning of life is intertwined with the essence of reality and the ever-present force of fear. Life's purpose is to live, to transform non-reality into reality, and to perpetuate existence. Through understanding this delicate balance, we can appreciate the profound significance of our own lives and the reality we inhabit.

Chapter 12.4: Deriving the God Particle

It is from whence we derive the god particle. Every species harbors a deep-seated fear of death and an intrinsic need to exist and recreate life. As soon as a species attains consciousness, it instinctively creates a mediator between existence and nonexistence—a force that will ensure its presence even if it momentarily slips into nonexistence.

Given our understanding of "time," it becomes evident that one could make a compelling case for the existence of a god. One might posit that God exists within a hypothesis or the Cartesian coordination system, where all points converge at the center in a singularity—a point where time, motion, and non-reality coexist. This realm is akin to the imagination, a space that is not real yet feels profoundly real. Like a god, this force may not have a tangible existence but exerts a significant influence on reality.

The Concept of the God Particle and Consciousness

The god particle, in this context, represents the mediator between existence and nonexistence. It embodies the force that sustains life and reality. This particle, much like an element of consciousness, ensures that the transition between non-reality and reality is seamless and continuous.

Chapter 12.5: The Role of Imagination in Reality

Imagination is a unique space where the boundaries between reality and non-reality blur. It is a place where ideas and concepts can exist without physical form, yet they hold the potential to shape reality. Just like the notion of a god, imagination is not bound by the physical constraints of existence. It is a realm where possibilities are infinite, and reality is malleable.

Imagination allows us to conceptualize and create, bridging the gap between what is and what could be. It is through imagination that we can hypothesize the existence of forces like the god particle, entities that may not be physically real but have a profound impact on our perception of reality.

The Interplay Between Imagination and Reality

The relationship between imagination and reality is dynamic and reciprocal. While imagination fuels our understanding of the universe and the possibilities within it, reality provides the framework within which these ideas can manifest. The god particle, as a mediator, operates within this interplay, influencing both realms and ensuring the continuity of existence.

Chapter 12.6: The Influence of Non-Physical Entities on Reality

Entities that exist within the realm of imagination or hypothesis, such as the god particle or a conceptual god, may not be physically real but can still affect reality. These non-physical entities represent the underlying forces and principles that govern the universe. They are the drivers of existence, the architects of reality, and the sustainers of life.

Just as a god may not have a tangible form but influences belief, morality, and actions, the god particle represents the abstract principles that shape the universe. It embodies the fundamental laws of existence, the interplay of time, motion, and non-reality, and the perpetual cycle of creation and transformation.

Conclusion: Embracing the Infinite Potential of Existence

The exploration of concepts like the god particle and the role of imagination in shaping reality allows us to appreciate the complexity and interconnectedness of existence. By acknowledging the influence of non-physical entities, we gain a deeper understanding of the forces that govern the universe. The meaning of life extends beyond mere physical existence; it encompasses the infinite potential of reality, the power of imagination, and the continuous dance between being and non-being.

Through this lens, we embrace the profound nature of existence, recognizing that our lives are part of a larger, ever-evolving cosmic narrative. We are both creators and creations, participants in the grand symphony of reality, and witnesses to the infinite possibilities that lie within the imagination.

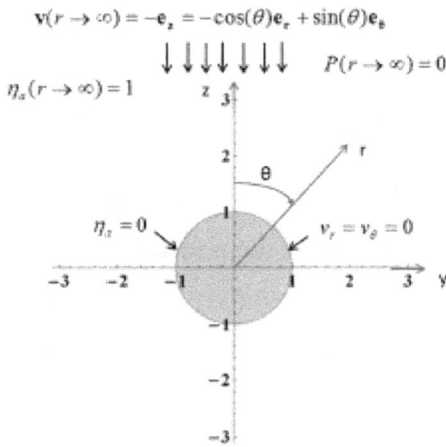

$$\mathbf{v}(r \to \infty) = -\mathbf{e}_z = -\cos(\theta)\mathbf{e}_r + \sin(\theta)\mathbf{e}_\theta$$

$$\eta_a(r \to \infty) = 1 \qquad P(r \to \infty) = 0$$

$$\eta_{,r} = 0 \qquad v_r = v_\theta = 0$$

Chapter 12.5: The Convergence of Reality and Non-Reality

We call this area where reality and non-reality meet "god." This is the crossroads between life and death, the force that maintains the delicate balance between non-reality and reality. Much like the Christians' concept of a mediator, such as the Bible or the God of the Koran, we anthropomorphize this intersection. We attribute to it characteristics and a semblance of consciousness, often imagining it as a being that mirrors human traits, despite it being only "half-real."

In our minds, this entity might display anger, hatred, love, and lust, similar to humans, though it lacks many components of humanity. This anthropomorphic view helps us make sense of the abstract forces that govern existence, yet it poses challenges for some, such as reconciling with the literal interpretations found in religious texts like the Bible.

Questioning the Nature of Light and Reality

This perspective on "god" also challenges conventional views on the role of light in the universe. The speed of light is often linked to concepts of time travel, but light is merely a construct of reality, not a fundamental function. Reality does not require light to exist. All existence ultimately converges at a single point in nothingness, and it is the very nature of being actual that dictates reality.

Chapter 12.6: The Infinite Cycle of Existence

In this framework, heaven can be seen as the opportunity to be reprogrammed into another form of reality. The molecules that once made up your heart muscle could now fuel the sun, contributing to its radiant energy. This transformation illustrates how each instance of existence creates new possibilities, continuously reshaping the universe.

The Role of Mediators in Existence

These mediators, or "half-real" beings, serve as the bridges between existence and non-existence. They ensure the continuity of reality, guiding the transformation of molecules and moments across different states of being. This process is not bound by the physical constraints of light or time but is governed by the fundamental principles of reality.

The Nature of Being and Transformation

Our essence, or "instants of existence," constantly evolves, contributing to the cosmic cycle. This transformation is akin to reprogramming— our molecules shift roles and purposes, perpetuating the cycle of existence. This perspective allows us to see death not as an end but as a transition to a different state of being.

Heaven as Reprogramming

In this view, heaven is not a distant, static realm but a dynamic state of reprogramming and renewal. It is the ultimate expression of the continuity of existence. Your essence, once part of your physical form, now fuels the cosmos, creating more possibilities and contributing to the endless cycle of life and reality.

Chapter 12.7: Embracing the Complexity of Existence

The meaning of life extends beyond mere physical existence. It is about embracing the complexity of reality and the continuous interplay between being and non-being. Our lives are part of a larger narrative, one that is ever-evolving and infinitely interconnected.

Conclusion: The Eternal Dance of Reality and Non-Reality

By understanding the convergence of reality and non-reality, we gain a deeper appreciation for the intricate forces that shape our existence. These mediators, whether conceptualized as gods or abstract forces, play a crucial role in maintaining the balance and continuity of the universe. Our essence, forever reprogramming and transforming, contributes to the grand symphony of existence, ensuring that reality remains vibrant, dynamic, and perpetually alive.

Challenging our perception of reality.

Chapter 13.1: The Nature of Nothing, Reality, and Infinity

What is meant by nothing and real and infinite? These concepts may sound like a divine spark, as if they carry a mystical quality. The Universe often behaves like an illusion, challenging our perception of reality. Let's delve into the mechanics behind this phenomenon.

When we investigate nothingness, we might find a tiny opening that we call reality. However, this reality seems to disappear and reappear. Imagine this opening expanding and pushing outward, then collapsing and spinning as nothingness spirals into infinity to close. Infinity always tends to zero, but if zero doesn't exist, it must be considered a point. Is zero even a real number, or is it simply an axiom where reality and non-reality intersect?

Something blocks this point, making it real, yet the point doesn't enter the imaginary singularity. As infinity tries to close this point, the point disappears with changes in the vector due to linear movement. Infinity continues to close in on the point until the next cycle, and the process switches back and forth. In the end, infinity never quite reaches zero because reality disappears somewhere along the way. Infinity may approach zero, but it will never actually reach it. This ongoing process is what we refer to as the divine spark.

Out of nowhere, someone might declare the existence of a zero hole. However, a zero hole, in essence, is no hole at all. Infinity spirals in an attempt to close a zero hole, revealing another illusion of the Universe: singularity.

Chapter 13.2: Understanding Singularity and Illusion

If you have followed my theory thus far, you will better grasp the illusion of the Universe. Investigating singularity reveals it as an infinite point within a singularity, but this is merely an illusion—much like a magician's sleight of hand. The speed of the hand creates the illusion, making one point appear as many.

Singularity represents a point into nothingness and is purely imaginary. It does not exist in a tangible sense. When reality tries to re-enter singularity, it fails because singularity doesn't exist. With no zero holes to accommodate reality, the motion of reality follows a vector, rotating positions and diminishing momentum as it attempts to enter singularity. Infinity spirals into the zero hole, and reality tries to enter at different vectors with diminishing momentum.

As reality tries to re-enter, it appears everywhere simultaneously at different locations due to instantaneous speed. This gives the illusion of multiple points, but it is actually one point attempting to re-enter singularity at different vectors and diminished momentum.

Chapter 13.3: The Infinite Illusion

If you examine infinity, you may see the illusion of an infinite point incursion into nothingness. However, this is merely one point manifesting at different vectors with diminished momentum. The diminished momentum causes reality to appear close and far as it increases and decreases.

This cycle creates a push and pull dynamic within the Universe. One moment, reality is tangible; the next, it slips into non-existence. This is why we often perceive the Universe as an illusion. We continuously spiral from nothingness back to reality, creating the illusion of an ever-changing cosmos.

Chapter 13.4: Movement, Time, and the Birth of the Universe

In the beginning, there was only movement and time—a collection of instants of reality. Each instant had motion and time. The Universe became populated with a single point that appeared everywhere as it attempted to re-enter singularity. This point manifested as multiple points, creating the illusion of plurality. It tried to renter singularity across vectors, diminishing momentum within nothingness.

The emergence of the Universe can be attributed to this process. Two clocks exist: an internal clock created by the Universe and a cosmic clock dictated by the appearance and disappearance of reality. Initially, there was no light or heat, only movement and time. As this singular point began to organize, it shaped the Universe as we know it today.

These imaginary points exert a force of singularity even when they come close to each other. The spin keeps them apart, creating gravity within the known Universe.

Conclusion: The Illusion of the Infinite

The Universe, with its illusions and realities, operates in a delicate balance of movement, time, and singularity. The interplay of these forces creates the dynamic cosmos we inhabit. Understanding the mechanics behind these illusions offers us a deeper appreciation of the intricate dance between nothingness, reality, and infinity. Through this lens, we can embrace the beauty and complexity of existence, recognizing that what we perceive as reality is but a fleeting moment in the infinite tapestry of the Universe.

Chapter 13.5: Cosmic Gravity and the Genesis of Matter

Cosmic gravity, governed by singularity, brings everything to a single point. Instant realities require singularity, and the rotation of these realities repels them apart, generating heat. This process marks the birth of suns, followed by planets and eventually organic matter.

The Dawn of Conscious Life

Conscious life began with the need for navigation. In its earliest form, life lacked self-propulsion, relying instead on environmental forces for movement—much like inanimate objects do. Only life forms possess the ability for self-propulsion. Objects in the Universe move due to external forces: wind blows under governing forces, and planetary movements and the sun's path follow similar principles. However, life evolved to navigate these forces independently.

Primitive movements were initially rudimentary, driven by environmental factors. Over time, sensory systems developed to control and refine these movements. As navigation became more sophisticated and sensory input improved, consciousness emerged. This higher level of awareness allowed for the prediction and transmission of information beyond genetic codes.

Chapter 13.6: The Mechanics of Movement in Reality

What is movement in this reality? Movement in this reality is the summation of all the different instants of reality through which you traverse. Each instant of reality encompasses motion, time, and vector. Consequently, time is the accumulation of all the realities you move through, giving us the concept of distance. The further you travel across realities, the further you are from your starting point.

There is a cost to movement: you lose some of your instant reality, which we perceive as aging. Before the advent of organic life, movement across the Universe was restricted by forces that pushed and pulled. The direction of movement across time is influenced by where unbalanced equations exist within the vector.

Chapter 13.7: The Balance of Forces and Self-Propulsion

Consider standing still: all forces and equations are balanced, meaning movements cancel each other out. Internal rotations within an object are balanced, with no push or pull. When a force acts on an object, it unbalances the equation in favor of the force's direction.

Self-propelled movement means not relying on external forces for movement across the Universe. It involves balancing the equation of instant realities in the direction desired by the conscious mind. Cosmic time is a fabric of this Universe, dictating the speed at which reality exists. This new Universe has its own set of time and rules.

Chapter 13.8: Time and the Universe

Time in the Universe varies for each object according to size and the forces acting upon it. In this reality, the more area an object covers, the more instants of reality it encompasses. A larger equation requires greater force to unbalance it and move in any direction. As a result, time increases for larger objects due to their greater time and motion. The potential energy in a large object is immense, explaining why smaller objects require less force to move—they have fewer instants of reality.

Chapter 13.9: The Cosmic Dance of Reality

The Universe is a complex interplay of forces, movements, and realities. The balance and unbalance of equations drive the motion of objects and life forms. Self-propulsion, a defining characteristic of life, allows organisms to navigate and adapt to these forces.

In this cosmic dance, the principles of cosmic time and the nature of reality shape existence. Understanding these mechanics offers insights into the fundamental workings of the Universe, revealing the interconnectedness of all things.

Conclusion: Embracing the Complexity of Existence

The Universe, with its intricate balance of forces and realities, operates in a dynamic and ever-evolving manner. By exploring the mechanics of movement, time, and gravity, we gain a deeper appreciation for the complexity and beauty of existence. This understanding allows us to embrace our place in the cosmos and recognize the profound interconnectedness of all life and matter. Through this lens, we can appreciate the delicate dance of reality and the forces that shape our world.

What is nothing in the scientific world?

Chapter 14.1: The Scientific Definition of "Nothing"

What is nothing in the scientific world? Nothingness can be perceived as eternal or timeless, simple, empty, plain, quiet, or even perfectly symmetrical—in which case, it is devoid of existence. Alternatively, nothingness might be beyond all existential description, implying it never was simple and never will be simple.

The Universe exists because it can exist. Before the Universe, there was nothing. Yet, the potential for nothingness to transform into the Universe existed. Since this possibility was present, it was bound to happen eventually, and so, here we are.

We must conclude that there is no divine entity orchestrating this process, for every probability has been accounted for.

Chapter 14.2: The Concept of Nonexistence

When we say something doesn't exist, it implies that certain entities, like unicorns or dragons, are not part of our reality. Magic carpets don't exist, at least on this planet and at this point in history. We can deny the direct existence of such things based on our current knowledge, though we could be wrong.

The absence of certain entities is merely due to the way matter and nature's forces are arranged. The Universe still exists without magical unicorns. Similarly, square circles don't have a place in reality because the rules of our universe establish what is meaningful. Things that lack meaning do not exist.

Chapter 14.3: Relative Existence in Time and Space

Consider a simple example: when we say milk doesn't exist in the refrigerator, we're stating that milk is not in that specific location at that specific time. This concept applies universally. According to a fundamental principle of time and space, things exist in locations relative to one another. If we remove these separate areas in time, everything exists in the same space. As physicist John Wheeler put it, "Time keeps everything from happening simultaneously."

Chapter 14.4: The Ambiguity of "Nothing"

In mathematics and physics, "nothing" does not have a clear designation. Like in many other fields, "nothing" is a general term with a vague definition, often used ambiguously. Scientists typically view nothing as being simpler than the Universe.

Chapter 14.5: The Emergence of the Universe from Nothing

The transition from nothing to the Universe is a profound concept. Before the Universe came into existence, there was a void—an absolute nothingness. This void, however, contained the potential for existence. This inherent potential led to the eventual manifestation of the Universe.

Chapter 14.6: The Illusion of Nonexistence

The idea of nonexistence can be challenging to grasp. Consider the notion of unicorns or dragons: they do not exist in our current reality, but this does not mean they are inherently impossible. Rather, our current understanding and arrangement of matter do not support their existence.

Similarly, the concept of "nothing" is an illusion. It is not a tangible state but a relative term. In the grand scheme of the Universe, nothingness is simply a state that precedes the potential for existence.

Conclusion: Understanding Nothing and Everything

The Universe's existence and the concept of nothingness are intertwined. By understanding the scientific perspective on nothing, we gain a deeper appreciation for the complexity and beauty of existence. Nothingness, with its potential for transformation, serves as the foundation for the Universe and all that it encompasses. Through this lens, we recognize that existence is a dynamic and ever-evolving phenomenon, shaped by the interplay of forces and the potential within nothingness.

Chapter 14.7: The Paradox of Nothingness

The astronomer David Darling remarks, "Nothing could be simpler than nothing -- so why is there something instead?" According to Occam's razor, the simplest answer is usually the correct one. When considering whether a universe should exist versus nothing at all, it seems logical that nothing would be more probable. However, this argument hinges on the assumption that nothingness is inherently simple.

Simplicity is a quality associated with uniformity, like a white canvas. Expecting nothingness to be simple is one of the best examples of how we mistakenly equate absolute nothingness with nonexistence.

Chapter 14.8: The Flaw in Nothingness

We might imagine nothingness as the default state before the Big Bang. However, nothingness has an inherent fault that allows for fluctuations in the void, causing something to emerge from nothing. Nothingness that doesn't exist cannot fluctuate or fracture, as there is nothing to sustain damage. In the absence of a universe, there is no time, change, or first cause. A first cause must be found within the initial stages of the Universe.

When we imagine absolute nothingness as a possibility, as the default state preceding the Universe, we inadvertently deny the possibility of our present existence. The concept of nonexistence attempts to specify the absence of existence and seems to work because other denials of existence achieve their purpose.

Chapter 14.9: The Void and Nonexistence

The word nonexistence aims to define the absence of existence. It appears effective because, in physical reality, when something doesn't exist, there is a void. However, the void left behind, such as when we say there is nothing in the refrigerator, is absolute, not nonexistent.

Chapter 15.1: The Illusion of Simplicity in Nothingness

Nothingness is often considered simpler than the Universe. This expectation arises from the belief that absolute nothingness possesses simplicity and uniformity. However, this perception is flawed. Absolute nothingness is not merely the absence of things but a state devoid of potential and fluctuation. The inherent instability of nothingness leads to the emergence of something.

Chapter 15.2: The Emergence of the Universe from Fluctuations

The idea that the Universe emerged from fluctuations in the void challenges our understanding of nothingness. These fluctuations indicate that nothingness is not a stable state. The inherent fault in nothingness allows for the creation of something, leading to the birth of the Universe.

Chapter 15.3: The Paradox of Nonexistence and Reality

Nonexistence attempts to define the absence of reality. However, this definition is problematic because it implies a void that is not truly empty. The void created by the absence of something, like the emptiness of a refrigerator, is absolute but not nonexistent. This paradox highlights the complex relationship between nothingness and reality.

Conclusion: The Complexity of Nothingness

Understanding nothingness reveals the inherent paradoxes and complexities of existence. The idea that the Universe emerged from fluctuations in nothingness challenges our perception of simplicity and uniformity. By exploring these concepts, we gain a deeper appreciation for the profound nature of reality and the intricate balance between nothingness and existence. This understanding allows us to contemplate the mysteries of the Universe and our place within it, recognizing that existence is a dynamic and ever-evolving phenomenon shaped by the interplay of forces and potential within nothingness.

As a result, we wonder why we exist instead of nothing because our thoughts default to a vague understanding of nothing and nonexistence. We fail to recognize that a universe must live, for nothing is not an alternative since nonexistence cannot be mindful. Nonexistence should seem to us to be impossible, and being should appear to us to be inevitable and natural.

As all forms become unified, they become a complete form. The individual states are lost, given to the whole structure, which we see as formless or neutral because we only see the many forms, things, objects, and imbalances. What constructs, and so we know the uniformity, the wholeness, and perfect symmetry of space as nothing in comparison. Men are not used to relating to any part of the world when it is unified and one.

"The Mechanics of Motion"?

Chapter 16.1: Time Travel and Understanding Motion

Before we can grasp the concept of time travel, we must first understand motion. Cosmic time differs from universal time. Cosmic time is measured by an instant of reality—the appearance and disappearance of reality at a given point in time. Universal time, on the other hand, is the summation of these cosmic instants across space.

Understanding Motion

People often think of motion merely as the movement of limbs, but it is far more complex. Every instant of reality involves motion. The reference we use to measure time is irrelevant because it is a social construct, born from human imagination. This is why traditional timekeeping methods, based on celestial movements, have their limitations.

Time and motion are specific to each object. Time behaves differently for large objects compared to small ones: the bigger you are, the more time you have; the smaller you are, the less time you have.

16.2: The Mechanics of Movement

An object moves through available space, navigating a gauntlet of time and motion. Movement in time is penalized by the gain and loss of time and motion. When an object moves through space, it encounters another instant of reality. Some instants of reality are removed or added to the object—this interaction is what we call friction.

Consider traveling 30 miles in one hour. What happens in that fixed point is the summation of all movement and time within the object. If you add, multiply, subtract, and divide all the movement and time, the net result is 30 miles in one hour. Larger objects require more time and motion for displacement.

16.3: The Instant of Reality

Think of each instant of reality as a tiny battery, fueled by motion and time. As you move through instants of reality, you gain and lose time and motion. The true measure of time and movement in the Universe depends on the quantity you have, not the rotational motion of the sun.

Scientists attempt to calculate time through space for time travel, but they mistakenly treat time as a linear concept, measured by the sun's rotation. In truth, real-time exists as individual pockets. Unless a force acting on an instant of reality can make all-time linear, the concept of linear time is flawed.

16.4: The Absurdity of Time Travel

The idea of time travel is riddled with paradoxes. The Universe creates points in space, forming imaginary points as it strives to return to singularity. To travel back in time, one would have to reconstruct all these imaginary points back to their previous frames, assuming the Universe even stores these frames—which is highly doubtful.

16.5: Reconstructing Reality

How can we find an imaginary point to recreate the past image? How do we reverse instants of reality backward in time? These questions highlight the implausibility of traditional time travel notions. The

Universe's frame-by-frame existence makes it nearly impossible to revisit previous states without an unfathomable reconstruction of every instant of reality.

16.6: The Limitations of Linear Time

Linear time is an oversimplification. It ignores the complex interactions and independent existence of each instant of reality. Just as a powerful magnet might polarize time into a linear path, imagining time travel requires a similarly impossible force to align all moments into a coherent, reversible sequence.

Conclusion: The Realities of Time and Motion

Understanding motion and time as distinct yet intertwined forces is crucial for any meaningful discussion of time travel. The reality is that each instant of time and motion is unique and independent. The Universe's complexity defies simple explanations, making traditional concepts of time travel more of a fantastical idea than a feasible reality. By acknowledging the intricate nature of time and motion, we can appreciate the true depth of the cosmos and the limitations of our current understanding.

Chapter 16.7: The Interdependence of Time and Motion

Time and motion exist together in an instant of reality. The more time and motion you possess, the faster you can travel. On a dense planet, you may live longer than on a less dense one, as your movement is restricted but you have more time.

16.8: The Challenge of Moving Through Space

Moving through space is difficult due to the interaction of time and motion. Time depends on the number of instances of reality you experience. Advanced technology allows faster movement, but friction becomes a limiting factor. As instant reality spreads apart, the energy required to move through space is limited, resulting in a floating sensation. Movement is possible in all directions, but energy constraints meantime and motion are reduced, leading to a shorter lifespan in space.

16.9: The Effect of Cosmic Scale on Time

In space, time is influenced by the quantity of instant reality. For instance, compared to the sun, there is very little time in space. The sun, with its vast mass, possesses much more time, causing a perception of time subtraction as the distance from the sun increases. This is due to the decreasing mass difference over time.

16.10: The Illusion of Time and Motion

When you subtract time and motion from available space, they diminish not for the object but for its surroundings. The object's time and motion remain constant, while the reality around it alters. This creates the illusion of changes in size or time. For example, removing an object from a solar system with 100 instants of time affects the surrounding reality more than the object itself.

16.11: The Loop of Cosmic Time

The Universe operates through illusions. Cosmic time and universal time are interconnected yet distinct. Cosmic time measures the appearance and disappearance of reality, forming a loop. Each loop is divided into 360 angles or vectors, with each vector containing an infinite point. Summing up these instants of reality gives us our known universe time. The summation of these cycles creates one complete circle or time frame. Our perception of time is based on how much we use within this cycle.

16.12: The Economy of Time and Motion

As we move through space, we buy, exchange, and sell time. Time passes slowly in space because there is little to exchange. This explains the long duration required to traverse the Universe. If rich fuel with abundant time and motion is found, faster travel becomes possible. When an object leaves a place, it subtracts from reality, creating the illusion of disappearance or backward movement in time.

16.13: Determining True Age

To determine your true age, you must sum all your time and motion. The result represents your true age, encompassing all movement and time over your lifetime. This cosmic age differs from solar age, reflecting the total accumulation of time and motion.

16.4: The Currency of Energy

To move through space and time, energy serves as currency. Energy is exchanged like a commodity as one navigates the cosmos. The availability of energy determines the feasibility of travel and interaction with reality.

Conclusion: The Complex Dance of Time and Motion

The interdependence of time and motion shapes our understanding of reality. Through this complex dance, we navigate the cosmos, continually exchanging energy to move through space and time. By comprehending this interplay, we gain insight into the profound mechanics of the Universe and our place within it

Chapter 17.6: The Enigma of Black Holes and Time Travel

It begs one to wonder what happens if a person gets sucked into a black hole. Could this be the only possible way to travel back in time within our Universe? Black holes represent where infinity begins and ends, where time and motion converge. They are the imaginary singularities in space and time, offering a glimpse into the origins of time and motion.

The creation of the Universe involved the thrust of a backward force to negative spin or negative time, followed by a forward force to positive spin or positive time. This push-pull dynamic caused rotation in opposing directions, creating both negative and positive time. As reality appears, rotation moves forward; as reality disappears, rotation reverses. Time progresses forward, but when time and motion return to zero, they rotate backward, creating a dual-clock system: one clock moving forward and another moving backward. Time generates progression, and time erases regression. If the last frame is saved, time travel on a galactic scale becomes possible.

17.7: The Mechanics of Black Hole Interaction

To understand the potential outcomes of entering a black hole, we need to consider the mechanics involved. Let's explore the endless possibilities:

Destruction into Infinity: The first possibility is that a person entering a black hole is destroyed, molecule by molecule, into infinity. This process would be irreversible, as the intense gravitational forces would disassemble matter into its fundamental components.

Reprogramming of Instant Reality: The second possibility is that the black hole reprograms the instant reality. In this scenario, the instants of reality that once constituted your heart muscle, keeping blood pumping through your veins, could be repurposed to become part of the sun's nuclear fission. This new reality highlights the continuous cycle of reprogramming and transformation.

Repetition of Entry and Exit: The third possibility suggests a cyclical return to the same point. Once you reach the center of the black hole, you might return to the exact moment you left, only to be sucked in again and repeat the process. This theory implies a loop, where you are continuously drawn back to the same point in space and time.

17.8: The Sling-Back Effect

Another theory posits that entering a black hole increases speed toward the center point, resulting in a sling-back effect. As you go from 0 to - 360, momentum decreases, causing a slot effect where energy diminishes. In this state, the rotation of time spins the instants of reality, scrambling to find empty slots in the spinning disk. These slots are filled until momentum decreases to zero, resulting in a loss of speed.

As reality loses speed, the disk spins in each rotational cycle, dropping instances of existence into tracks with no empty spots. The spinning continues until it becomes zero. The swing forward on the positive side also loses momentum until it returns to the slots. This intricate dance of time and motion within the black hole creates a complex interaction of forces and realities.

17.9: The Endless Possibilities

Black holes present a multitude of possibilities, each highlighting the intricate relationship between time, motion, and reality. These cosmic phenomena challenge our understanding and push the boundaries of what we know about the Universe.

Conclusion: Embracing the Mystery of Black Holes

Black holes remain one of the most enigmatic and fascinating phenomena in the cosmos. By exploring the potential outcomes of entering a black hole, we gain insight into the complexities of time, motion, and reality. These theories, while speculative, open the door to endless possibilities and encourage us to continue seeking answers to the mysteries of the Universe. Through this exploration, we deepen our appreciation for the profound nature of existence and the intricate dance of forces that shape our reality.

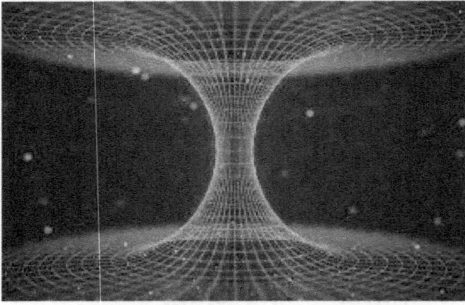

Turns out Darwin was right

Chapter 18.1: Evolution Over Time

Turns out Darwin was right; everything changed over time. Charles Robert Darwin, an English naturalist, geologist, and biologist, is best known for his contributions to the science of evolution. His proposition that all species of life have descended from common ancestors is now widely accepted as a fundamental scientific concept. To better understand Darwin's theory, let's rewind the clock to the first stroke of existence.

18.2: The Birth of Reality

At that primordial moment, an instant of reality first appeared. Time existed, but motion had not yet emerged. This means that at this moment, you could be everywhere simultaneously without any movement to separate time into vectors. How could you be everywhere at once? By existing as a dot. The motion began when that dot disappeared and reappeared. This marked the birth of singularity—a point in the nothingness where many dots appeared.

18.3: The Formation of Singularity

Singularity comes into existence when you examine a dot and see a multitude of dots trying to cover space. This might seem abstract but think of it in terms of a computer monitor. When creating an image on a screen, the first step is to establish a Cartesian field by dividing the monitor into a plane, then subdividing that plane into quadrants filled with pixels. Within each pixel lies the essence of singularity—an imaginary construct that represents the dot.

18.4: The Evolutionary Process

Fast forward through billions of years, and we see the intricate dance of evolution. Life forms emerged, adapted, and evolved. Darwin's theory proposed that through natural selection, species change over time, giving rise to the diverse array of life we see today. This process, driven by genetic variation and environmental pressures, shapes the survival and reproduction of organisms.

18.5: The Interplay of Time and Motion

Time and motion are intertwined in the fabric of reality. As motion began, time started to differentiate into vectors. This dynamic interplay allowed for the emergence of complexity and diversity in life forms. Evolution is a testament to this intricate relationship between time and motion, where each instant of reality contributes to the unfolding narrative of life.

18.6: Understanding Evolution Through Modern Science

Modern science has provided us with tools to understand and validate Darwin's theory. Genetic research, fossil records, and comparative anatomy all support the idea that life has evolved through common descent. The study of DNA has revealed the genetic links between species, showcasing the gradual changes that have occurred over millennia.

18.7: The Impact of Darwin's Theory

Darwin's theory of evolution revolutionized our understanding of life. It provided a framework to explain the diversity of species and the mechanisms of adaptation and survival. This scientific perspective challenged existing beliefs and paved the way for new discoveries in biology, genetics, and anthropology.

Conclusion: The Endless Journey of Evolution

Evolution is an ongoing process, a journey that continues to shape the living world. Darwin's insights laid the foundation for our understanding of this journey, revealing the interconnectedness of all life. By recognizing the profound impact of time and motion on the evolution of species, we gain a deeper appreciation for the complexity and beauty of the natural world. Through this lens, we can marvel at the ever-changing tapestry of life and our place within it.

Chapter 18.8: The Dot Within Dots

You look at the dot; inside the dot, bunches of dots are trying to fill a plane. And within each of those dots, more dots are trying to fill space. What is the plane the dot attempts to fill but an imaginary time and space of nothingness? This mirrors the thought or instant of reality—a representation of consciousness when you recreate a mental image of the natural world in your mind. Higher consciousness manipulates and predicts the movement of objects.

By investigating the dot, you create a singularity or a thought. Both are imaginary yet real, marking the beginning of consciousness. Imagine the first organic matter trying to navigate. Initially, navigation relied on simple sensors: if something was detected, a reaction occurred, prompting movement. As sensors became more complex and guidance systems more sophisticated, a synthetic version of the natural world formed in the organism's mind for navigation.

18.9: The Emergence of Conscious Thought

When your mind fills with dots forming quadrants, that moment signifies singularity or thought. When organic matter recreates an idea or singularity, it creates both a real and imaginary construct. This is the foundation of higher cognitive processes.

The stage of energy

Chapter 19.1: Time, Motion, and Change

Returning to Darwin's theory, we acknowledge that everything changes over time, which is inevitable. In the initial instance of reality, time and motion existed simultaneously as a dot. Visualize the Cartesian coordinating system: the center dot represents time and motion. Over time, these two perpendicular forces separated. As the dot disappeared and reappeared, time evolved, and motion separated from it, creating a dynamic interplay.

Time emerged as circular motion, while movement appeared linear. The dot filled the void of nothingness, existing everywhere simultaneously. Motion became the vector of time, with its strength or approximation determining the singularity and rotation of each instant of reality.

19.2: Understanding Time's Transformation

As movement changes, so does the vector of a moment of truth. An instant of existence does not always return to the same position. Protons, neutrons, and electrons change configurations due to interchangeable spins, leading to transformations over time.

19.3: The Nature of Energy and Matter

Before delving into further changes, we must understand energy's significance. Matter and energy are interchangeable; matter is energy, and energy is matter. They share the same fundamental dot. The transformation between energy and matter involves stages of an atom. At the outer covalent and inner nucleus bond levels, atoms undergo changes. When an atom approaches singularity, the proton and neutron merge, entering a third stage—energy.

19.4: The Stage of Energy

In this stage, the spin rotation does not complete a full cycle; instead, it exhibits super speed with rapid back-and-forth motion. This supercharged instant of reality, called energy or forces, possesses such velocity that it can move objects within this imaginary world.

Conclusion: Embracing the Complexity of Reality

By understanding the interplay of dots, consciousness, time, motion, and energy, we gain insight into the profound nature of reality. Darwin's theory of change over time is an essential part of this understanding. As we explore these concepts, we appreciate the intricate dance of forces that shape our existence. Through this lens, we recognize the interconnectedness of all things and the dynamic evolution of the Universe, driven by the continuous interaction of time, motion, and energy.

Chapter 19.5: The Nature of Time and Energy

You can say that energy is half of time or supercharged time. Can time be divided in half? Can it be divided theoretically into quarters? This division of time results in variations in energy, with some being more potent than others, leading to the concept of unlimited energy or speed where time and motion converge. This brings us back to the beginning of time.

The Real Mechanism of Time Travel

If time travel exists, it would not rely on the speed of light but rather on the speed of the electron. Electrons move so swiftly that they can reverse time and motion back to a specific point. In the Cartesian field, time and motion are perpendicular. Imagine an instant of reality spinning so fast that it reduces the sector from 360 degrees to 180 degrees. Continuing this reduction results in a negative spin from 360 to -90, to -180, and back to -360 or zero.

Time travel using the speed of light is impossible unless it matches the speed of an instant reality. If light travels at the same speed as an instant of existence, it must not move quickly enough to reach a point where time and motion converge at zero.

19.6: The Concept of Superluminal Speed

There must be a speed faster than light. Light, at best, may only be one degree or 1/360 of this faster speed. If we multiply the speed of light by 360, we obtain the speed needed for time travel when speed and motion are unified.

Traveling at Superluminal Speed

Imagine traveling at such a speed that you exist simultaneously everywhere because motion ceases to exist. You would be everywhere at the same time, but as a dot. This dot forms the basis of the Universe, dependent on motion. Without motion, existence as we know it ceases, and we return to our natural state as a dot. Removing motion removes space, causing everything to exist as a single point.

19.7: The Reversal of Reality

Reversing that point from zero to negative changes the reality object's motion on a reverse rotation. This recreates the moment in reverse order because the point already exists. The only thing moving backward is the object created by that point. Therefore, time travel is possible but does not depend on the speed of light.

19.8: The Speed of Light and the Universe

The speed of light is not a cosmic function but a universe function. It is a construct of reality rather than part of it. Imaginary concepts cannot interact with truth. Light, being imaginary, creates the fictional Universe. The cosmic definition is based on reality, while morning is a substantive construct of the known Universe.

The Cosmic and Universal Clocks

The external clock of the cosmos can be altered to any unit of time, such as seconds. One instant of real-time equates to the internal clock of the Universe—the speed of light.

19.9: Unveiling the Mechanics of Time Travel

Understanding time travel involves recognizing that light is merely an internal clock for the Universe. The true mechanism lies in the speed beyond light—the instantaneous speed where time and motion merge. By grasping this concept, we can explore the possibilities of traveling through time and space.

Conclusion: The Infinite Dance of Time and Motion

Time travel hinges on the intricate relationship between time, motion, and energy. By comprehending the true nature of these forces, we can unlock the mysteries of the Universe. This exploration reveals that time travel is not about surpassing the speed of light but understanding the deeper mechanics of existence. Through this lens, we appreciate the endless dance of time and motion that shapes our reality and opens doors to new possibilities.

Chapter 20.0: Energy as a Commodity

We see energy as a desirable and valuable commodity for exchange. The more energy you have, the more valuable it becomes, depending on its magnitude. For instance, $100 is more desirable than $50, and 50 $100 bills are more desirable than 10 $10 bills.

Energy and Evolution

Darwin's theory of evolution illustrates how changes in instant reality, such as becoming a proton, neutron, electron, or energy, lead to lasting transformations. The rotational spin of these particles changes over time. When protons and neutrons merge, they cheat reality by diminishing the vector spin from a full circle to a half circle.

Over time, as you move through time and motion, interactions with your environment cause changes. Objects constantly exchange instants of reality with their surroundings, using energy for movement. This continuous exchange affects evolution. Beneficial changes allow species to survive, while detrimental changes may lead to extinction. Species can evolve to become unrecognizable from their ancestors or may completely change their configuration and makeup.

20.1: The Beginning of Consciousness

The onset of consciousness involves investigating singularity to recreate the natural world. As beings, we exist in a higher state of consciousness, manipulating objects using Boolean functions. A Boolean function assumes values from a two-element set, performing logical operations on binary inputs to produce a single binary output. This concept is fundamental in logic, Boolean algebra, and switching theory.

20.2: Higher States of Consciousness and Boolean Logic

Using a logic gate as an ideal model for computation, a logical operation is performed on binary inputs to produce a single binary output. Depending on the context, the term "logic gate" may refer to an ideal model with zero rise time and unlimited fan-out, or a non-ideal physical device. Compound logic gates such as AND-OR-Invert (AOI) and OR-AND-Invert (OAI) illustrate this complexity.

20.3: Mathematical Functions and Relationships

In mathematics, a function is a binary relation between two sets, associating each element of the first set with precisely one element of the second set. Typical examples include functions mapping integers to integers or natural numbers to real numbers.

The Role of Energy in Evolution

Energy plays a crucial role in evolution, acting as a driving force behind changes in species. As species adapt to their environments, their energy exchange and usage patterns evolve. This process shapes the survival and reproduction of organisms, ultimately influencing the course of evolution.

20.4: The Exchange and Transformation of Energy

Energy transformation is a constant process in the Universe. Matter and energy are interchangeable, with matter being a form of energy and vice versa. This interchangeability drives the dynamic nature of existence. At the atomic level, the outer covalent and inner nucleus bonds define an atom's structure. When an atom approaches singularity, protons and neutrons merge, entering a third stage—energy.

Energy and Atomic Structure

In this third stage, the spin rotation of particles does not complete a full cycle. Instead, it exhibits super speed with rapid back-and-forth motion. This supercharged instant of reality, known as energy or force, possesses such velocity that it can move objects within the imaginary world of singularity.

20.5: The Future of Energy and Consciousness

As our understanding of energy and consciousness evolves, we may unlock new possibilities for manipulating reality. The continuous exchange and transformation of energy will drive the progress of science and technology, shaping the future of human civilization.

Conclusion: Embracing the Complexity of Energy and Evolution

By understanding the interplay of energy, consciousness, and evolution, we gain insight into the profound nature of reality. Darwin's theory of change over time, combined with modern scientific concepts, reveals the intricate dance of forces that shape our existence. Through this lens, we recognize the interconnectedness of all things and the dynamic evolution of the Universe. This understanding allows us to appreciate the beauty and complexity of life, energy, and the cosmos, inspiring us to explore the endless possibilities that lie ahead.

Chapter 20.6: Boolean Functions and Their Significance

A Boolean function describes an algebraic expression consisting of binary variables, the constants 0 and 1, and logic operation symbols. For a given set of values of the binary variables involved, a Boolean function can have a value of 0 or 1. For example, the Boolean function can be defined in terms of three binary variables. The function equals 1 if and only if certain conditions are met.

Expressions and Truth Tables

Every Boolean function can be expressed as an algebraic expression or in a Truth Table. While a function may be expressed through various algebraic expressions due to logical equivalency, each function has a unique Truth Table. When transformed from an algebraic expression into a circuit diagram, a Boolean function is composed of logic gates connected in a particular structure. A function associates each element xx of set XX, the domain, with a single element yy of set YY, the codomain.

20.7: The Role of Boolean Functions in Consciousness

Once organic matter begins investigating the face of God or the singularity, it is alive and conscious. When this matter starts associating images and manipulating ideas to question higher concepts, it exists at a higher level of consciousness. You are alive because you try to recreate the world's image around you—only living entities can do that. The term "living" means having the ability to self-propel. The moment you can self-propel, you are alive.

20.8: Levels of Existence and Consciousness

The level of existence depends on your stage in the self-propel motion. Are you at the beginning stage, only detecting positions? Are you in the self-navigation stage, capable of recreating reality in your mind? Or are you at a level where you can manipulate objects to achieve results?

20.9: The Mechanics of Boolean Functions in Higher Consciousness

In the realm of higher consciousness, Boolean functions play a crucial role. Boolean logic provides a framework for understanding and manipulating binary variables, which are essential for computing and decision-making processes. As consciousness evolves, the complexity of these functions increases, allowing for more sophisticated interactions with reality.

Boolean Logic and Cognitive Processes

Boolean logic is foundational in cognitive processes, enabling organisms to process information, make decisions, and interact with their environment. The manipulation of binary variables through logic gates mirrors the way our brains process information at the neural level.

21.1: The Interplay of Energy and Boolean Functions

Energy and Boolean functions are interconnected in the evolution of consciousness. The efficient use of energy allows for the execution of complex Boolean operations, driving the advancement of cognitive abilities. As organisms develop more efficient ways to use energy, they can perform more intricate computations, leading to higher levels of consciousness.

21.2: The Evolution of Computational Thinking

The development of Boolean functions parallels the evolution of computational thinking in organisms. As species evolve, their ability to process information and solve problems using Boolean logic improves.

This advancement is evident in the complexity of neural networks and the sophistication of decision-making processes in higher organisms.

Boolean Functions in Artificial Intelligence

In the field of artificial intelligence, Boolean functions are fundamental. They enable machines to process data, make decisions, and perform tasks that require logical reasoning. The evolution of AI mirrors the evolutionary processes observed in natural organisms, with increasing computational power leading to more advanced capabilities.

Conclusion: The Infinite Possibilities of Boolean Logic

By understanding the role of Boolean functions in consciousness and computation, we gain insight into the profound nature of cognitive processes. The interplay of energy, logic, and evolution drives the continuous advancement of both natural and artificial intelligence. Through this exploration, we appreciate the intricate dance of forces shaping our reality, recognizing the endless possibilities that lie ahead. This understanding inspires us to explore the depths of consciousness and computation, unlocking new potential for growth and innovation.

Chapter 21.03: Dedekind Numbers

In the vast landscape of mathematics, certain numbers carry with them a sense of mystery and depth that beckons us to explore further. Dedekind numbers are among these intriguing figures, arising from the interplay between combinatorics, logic, and lattice theory. They capture the essence of monotone Boolean functions and reveal the complexity hidden within seemingly simple systems.

Understanding Boolean Functions

To appreciate Dedekind numbers, we must first delve into the world of Boolean functions. A Boolean function is a mathematical function that takes inputs of binary values—0 or 1, representing false or true—and produces a binary output. With n variables, there are 2^n possible input combinations. The number of distinct Boolean functions on n variables is 2^{2^n}, an exponential explosion showcasing binary logic's richness.

However, our focus narrows to a special class known as monotone Boolean functions. These functions possess a property of order preservation: if the input variables are increased (in the Boolean sense, where 1 > 0), the output does not decrease. Formally, for inputs x and y, if $x \leq y$, then $(x) \leq f(y)$. This monotonicity aligns with intuitive notions of accumulation or growth; turning more switches "on" does not result in a decrease in the overall system's output.

Defining Dedekind Numbers

A Dedekind number, denoted often as D(n), counts the total number of monotone Boolean functions of n variables. Calculating Dedekind numbers is an arduous task due to their rapid growth. The initial values are known:

- $D(0) = 2$
- $D(1) = 3$
- $D(2) = 6$
- $D(3) = 20$
- $D(4) = 168$
- $D(5) = 7,581$
- $D(6) = 7,828,354$

Beyond n = 6, the numbers become extraordinarily large, and exact values for higher n are challenging to compute.

Interpretation Through Set Systems

Another lens to view Dedekind numbers is through the concept of set systems. Consider the power set of an n-element set, which consists of all possible subsets. The number of subsets is 2^n. A monotone set system (or monotone family) is a collection of subsets that is closed under superset formation; if a set is in the family, all its supersets are also included.

The Dedekind number $D(n)$ then counts the number of such distinct monotone families. This perspective connects Dedekind numbers to the theory of antichains and partially ordered sets (posets), enriching their combinatorial significance.

The Geometry of Dedekind Numbers

Visualizing Dedekind numbers brings us to the realm of hypercubes. For n variables, imagine an n-dimensional cube where each vertex represents a unique combination of input variables—a binary string of length n. The edges connect vertices that differ by a single bit.

Monotone Boolean functions correspond to order-preserving mappings on this hypercube. The functions partition the cube into regions that reflect the function's output, and counting these partitions aligns with determining Dedekind numbers. This geometric interpretation aids in conceptualizing the abstract numerical growth.

Historical Context and Significance

Named after the German mathematician Richard Dedekind (1831–1916), Dedekind numbers honor his extensive work in abstract algebra and number theory. Dedekind's contributions laid foundational stones in rigorous mathematics, including concepts of rings, fields, and lattice structures.

The study of Dedekind numbers not only pays homage to his legacy but also continues to animate contemporary mathematical research. These numbers appear in discussions on computational complexity, the structure of free distributive lattices, and even theoretical computer science, where understanding logical functions is paramount.

Challenges in Computation

Calculating Dedekind numbers for higher values of n is notoriously difficult. The combinatorial explosion requires sophisticated algorithms and immense computational resources. For instance, determining D⑦ was a significant achievement that required advanced techniques and clever optimizations.

Research into Dedekind numbers often involves:

- Recursive Methods: Breaking down the problem into smaller, more manageable parts.
- Symmetry Exploitation: Leveraging the inherent symmetries in monotone functions to reduce computation.
- Parallel Computing: Utilizing multiple processors to handle the vast computations required.
- The difficulty in calculating these numbers underscores their complexity and the subtle intricacies of monotone functions.

Connections to Other Mathematical Concepts

Dedekind numbers intersect with various mathematical domains:

- Lattice Theory: The set of monotone Boolean functions forms a lattice under function composition, revealing rich algebraic structures.
- Combinatorics: Antichains and posets relate directly to Dedekind numbers, as they represent collections of elements with specific ordering properties.
- Logic and Computer Science: Understanding monotone functions aids in circuit design, optimization, and the analysis of algorithms.

These connections highlight the Dedekind numbers' role as a bridge between abstract theory and practical application.

Beyond the Numbers

While the numbers themselves are captivating, the exploration of the underlying concepts offers a deeper reward. Investigating Dedekind numbers encourages us to ponder questions such as:

- How do order and structure manifest in complex systems?
- What are the limits of computation in combinatorial problems?
- How can abstract mathematical ideas inform technological advancements?

Engaging with these questions not only satisfies intellectual curiosity but also propels innovation across disciplines.

As we reflect on Dedekind numbers, we're reminded of the profound depth that can emerge from simple beginnings. From the binary realm of zeros and ones arises a tapestry of mathematical wonder, weaving together logic, geometry, and algebra into a unified whole.

For those eager to delve further, consider exploring related topics:

- Free Distributive Lattices: Structures that generalize Boolean algebras and connect to Dedekind numbers.
- Antichain Enumeration: Understanding how antichains contribute to combinatorial counting problems.
- Computational Complexity: Examining the limits of algorithmic computation in combinatorics.

The journey through Dedekind numbers is but one path in the infinite expanse of mathematical exploration—a path that continues to inspire and challenge mathematicians and enthusiasts alike.

Abiogenesis – A brief history

Chapter 22.0: The Genesis of Life from Nonliving Matter

Even though Darwin focused on the origin of species, some scientists have extended the concept of evolution to abiogenesis, the origin of life from nonliving matter. In 1924, Russian biochemist Alexander Oparin proposed that living cells gradually arose from nonliving matter through chemical reactions. According to Oparin, gases present in the atmosphere of primitive Earth, when induced by lightning or other energy sources, reacted to form simple organic compounds. These compounds self-assembled into increasingly complex molecules, such as proteins, which eventually organized into living cells.

Chapter 22.1: The Miller-Urey Experiment

In 1953, Stanley Miller and Harold Urey tested Oparin's hypothesis by conducting an experiment to simulate the atmospheric conditions of primitive Earth. In their experiment, water was boiled into vapor at the bottom of a flask and then passed through an apparatus containing ammonia, methane, and hydrogen. The resulting mixture was subjected to a 50,000-volt spark, then cooled and collected in a trap. Upon examination, Miller and Urey discovered a collection of amino acids, the building blocks of life.

Chapter 22.2: Challenges of Abiogenesis

However, Miller's experiment did not accurately simulate primordial Earth's conditions. For instance, oxygen was present on early Earth, and oxygen inhibits the formation of organic compounds. Although we need oxygen to survive, our bodies have adaptations to manage it safely. In the 1950s, researchers assumed early Earth had very little oxygen, but geological evidence suggests substantial oxygen was present in Earth's earliest atmosphere. Using the gases now believed to have been present on early Earth would not produce amino acids.

Chapter 22.3: The Issue of Amino Acid Chirality

Even if Miller's experiment faithfully recreated early Earth's conditions, another significant difficulty arises: the chirality of amino acids. Amino acids exist as mirror isomers, with left-handed (L-form) and right-handed (D-form) versions. Living proteins comprise left-handed amino acids, but simulations like Miller's produce equal mixtures of both forms. Natural mechanisms produce amino acids in the same proportion of right- and left-handed forms. Even if a natural means were found to segregate the left-handed forms needed for life, it remains unclear how these L-form amino acids would become correctly ordered and linked to form proteins. The odds of obtaining a single protein from a primordial soup of exclusively L-form amino acids are slim.

Chapter 22.4: The Complexity of Forming Living Cells

Suppose a naturalistic mechanism was discovered to segregate left-handed amino acids, and a mystical soup could form proteins. Developing a living cell requires hundreds of specialized proteins to be precisely coordinated. Additionally, DNA, RNA, cell membranes, and various other chemical compounds must be produced and arranged correctly to perform their functions.

Chapter 22.5: The Search for Life's Origins

The quest to understand the origins of life continues. Scientists explore various hypotheses and conduct experiments to replicate the conditions that might have led to early life.

Chapter 22.6: Abiogenesis – Conclusion

The Challenge of Abiogenesis To transition from the conditions simulated in the Miller-Urey experiment to a living cell via unguided materialistic processes is to stack improbabilities upon improbabilities. Dean Kenyon aptly concludes, "It is an enormous problem how you could get together in one tiny, sub-microscopic volume of the primitive ocean all of the hundreds of different molecular components you would need for a self-replicating cycle to be established."

Chapter 22.7: RNA World – An Introduction

The RNA World Hypothesis The RNA world hypothesis offers a solution to origin-of-life researchers' challenges regarding the original information storage medium on the primitive Earth. DNA houses the cell's information necessary for folding proteins into the correct shapes critical for their respective functions. Every cellular and extracellular structure is constructed from proteins. Consequently, the information in DNA defines life's most fundamental operations and structures.

During cell replication, DNA and its stored information are copied and passed to daughter cells. Biochemical blueprints are conveyed to the next generation through DNA replication, resulting in two 'daughter' molecules identical to the 'parent' DNA molecule. One copy is complete; the two generated DNA molecules are distributed between the daughter cells produced during cell division.

Building proteins requires the genetic information in DNA, yet the data in DNA cannot be processed without many specific proteins and protein complexes. The mutual interdependence of DNA and proteins has been a significant stumbling block for Darwinian paradigms regarding life's origin since the mid-1980s. Origin-of-life researchers even refer to this conundrum as the chicken-and-egg paradox. Given that proteins are fundamental to how DNA replicates, DNA and proteins could not simultaneously arise from a primordial soup.

Chapter 22.8: RNA World – A Solution?

A Proposed Resolution: The RNA world hypothesis has been proposed as a resolution to this paradox. This model posits that RNA preceded DNA and proteins as the initial fundamental information storage medium. RNA can simultaneously store information (like DNA) and catalyze chemical reactions (like proteins). Thus, the RNA world eventually evolved into the DNA-protein world of contemporary biochemistry, with RNA now functioning as an intermediary between DNA and proteins.

While the RNA-world hypothesis sidesteps the need for an interdependent system of DNA and proteins in the earliest living system on paper, it appears untenable in practical terms. Numerous difficulties abound for the RNA world hypothesis.

Difficulties with RNA Formation: For instance, forming the first RNA molecule would have necessitated the prior emergence of smaller constituent molecules, including ribose sugar, phosphate molecules, and the four RNA nucleotide bases. However, both synthesizing and maintaining these essential RNA-building molecules (particularly ribose) and the nucleotide bases are profoundly problematic, if not impossible, under realistic prebiotic conditions.

Limited Enzymatic Properties: Another significant difficulty confronting proponents of the RNA-world hypothesis is that naturally occurring RNA molecules possess very few specific enzymatic properties of proteins. Ribozymes can perform a small handful of the thousands of functions performed by proteins.

The Transition to Modern Cellular Systems: The inability of RNA molecules to perform many of the functions of protein enzymes raises another concern regarding the tenability of the RNA-world paradigm. There is no plausible explanation for how primitive self-replicating RNA molecules could have transitioned into modern cellular systems that rely heavily on various proteins to process genetic information. Transitioning from a primitive replicator to a system for building the first proteins is a significant leap. Even if a system of ribozymes for building proteins had arisen from an RNA replicator, that system of molecules would still require information-rich templates for building specific proteins. There is no foreseeable account of the origin of that information.

Continuing the Investigation: Despite these challenges, the RNA world hypothesis remains a focal point in the ongoing investigation into life's origins. Researchers continue to explore and refine this hypothesis, hoping to uncover more plausible pathways that could have led to the emergence of life on the early Earth. As our understanding of biochemistry and prebiotic chemistry advances, we may yet discover new insights that bring us closer to solving the profound mystery of life's origins.

Chapter 22.9: RNA World – Conclusion

The Role of RNA in the Origin of Life: In summary, RNA is limited in its functional capacity, performing only a few minor roles. These roles often result from researchers intentionally engineering the RNA catalysts in question, further underscoring its limitations. Despite these challenges, the RNA world hypothesis continues to hold sway among most neo-Darwinians. They remain convinced that the RNA world must have existed, paving the way for the DNA-protein world that characterizes contemporary biochemistry.

Enduring Challenges and Paradoxes: The RNA world hypothesis, while providing a conceptual framework for the early stages of life's development, leaves several significant challenges unresolved. Chief among these is the enduring chicken-and-egg paradox. This paradox highlights the mutual dependency of DNA and proteins, raising the question of how such an interdependent system could have arisen in the first place.

Engineering and Evolution: The engineering required to demonstrate RNA's catalytic capabilities suggests that natural, unguided processes might not have sufficed to establish these functions initially. This engineered assistance points to the improbability of RNA alone solving the complexities of life's origins. Nevertheless, the hypothesis suggests a plausible evolutionary pathway wherein RNA, with its dual role as both genetic material and enzyme, gradually led to more complex systems involving DNA and proteins.

Continuing the Search for Answers: As research continues, the scientific community seeks to unravel these complexities and provide more concrete answers to the origins of life. The RNA world hypothesis remains a critical area of study, inspiring innovative experiments and new theories that may eventually bridge the gaps in our understanding.

Future Prospects and Theoretical Advances: Future research may provide new insights that either support or challenge the RNA world hypothesis. Advancements in our understanding of prebiotic chemistry, the discovery of new catalytic properties of RNA, or the identification of alternative pathways for life's origin could all contribute to refining or redefining this hypothesis. The quest to understand life's beginnings is ongoing and each discovery brings us closer to a comprehensive theory that unites the chemical and biological sciences in explaining how life arose on Earth.

The moment of creation!

Chapter 30.0: The Nature of Being Alive

Understanding Existence: What does it mean to be alive? To be alive, one must transition from potential to action, from existence to activity. In the vast cosmos, time is measured differently from our universe. The cosmos operates on a clock that sets a limit on speed—anything exceeding that speed ceases to exist. When entering our universe, one must obey certain boundaries, distinct from the cosmic laws. In this new reality, forces are governed by the cosmos, while energy serves as the prime mover of our universe. Forces drive the cosmos, akin to the movement of a piston in a two-stroke engine design.

In the cosmos, singularity is feasible because forces are stronger than energy. Motion lacks a vector—it is defined solely by time. Conversely, in our universe, time is linear, and the speed limit is the speed of light. Movement here is determined by the motion of an object, not by time and motion as in the cosmos. Motion has a vector or direction, and there are strict laws in place, including a speed limit. If a galaxy breaks this limit, it is expelled back to the cosmos as a singularity.

The Foundation of Life: To be alive in this universe, you must transition into activity. Symbolically, one might say you must sit at the right hand of a higher power to be granted life. The newly created object begins its daily operations, creating complex circuitry for navigation—typically the first completed circuitry. For effective navigation, the object must form a mental image of its surroundings. Being alive entails awareness of the world around you. How can you navigate if you are unaware of your surroundings? This awareness requires building sensors and control mechanisms to sense your environment. Once this mechanism is in place, the object is considered alive.

The Forbidden Knowledge: Becoming alive involves acquiring knowledge, akin to eating forbidden fruit. How do you consume this forbidden fruit or attain awareness of your natural surroundings? By forming a mental picture of your environment—its appearance, texture, and smell. The apparatus for consciousness has been created and now needs activation. Investigate the singularity and ask if you wish to exist. A higher power or singularity evaluates your readiness by examining if the necessary apparatus is installed. This evaluation marks the ceremony of transition.

The Laws of Existence: In this universe, you cannot exist as a singular entity. Violating this prime directive results in reverting to a singularity. We value this directive so much that we assign a numerical representation to the speed limit. Exceeding this speed violates our first prime directive and reverts you to a singularity. Additionally, you must accept that all instances of reality exist as multiple instances. The back-and-forth motion of time in the cosmos continues to spread the universe, creating space. A hole forms in the new universe, allowing motion to have a vector quantity and move in any direction. Motion in this universe is not just back and forth—it is an instant of reality or a dot. Time is linear, as motion depends on the flipping of pages and energy. You cannot be everywhere simultaneously; you exist as a dot, and being a dot means closing off other realities.

Creating a Mental Image: To create a mental image of reality in your mind, look into the dot. What do you see? Multiple dots spread all over—that is your new reality. You are ready to be alive when you realize that instant existence can exist as multiple dots. Look into the dot again. What do you see? Multiple instances of reality within that dot, mark the end of existence. Now you have an endpoint were reality transitions into nothingness. The multiple dots you see represent our reality. On the other side of the dot is an imaginary counterpart. Now that you understand the universe has an endpoint rather than an infinite point like the cosmos, you grasp the essence of being alive in this universe.

With this realization, you are prepared to embrace life, navigate the complexities of existence, and adhere to the laws that govern our universe. Welcome to the dynamic interplay of forces and energies that define what it means to be alive.

Chapter 30.1: The Nature of Imaginary Reality

Imaginary Constructs and the Universe: What you are looking at is imaginary, a copied image of reality. The universe has an endpoint because it extends into nothingness from the point of instant existence. Once you establish an endpoint, you can create a plan, intersect the quadrant, and fill the other side of reality with the imaginary part of reality. The two sides of reality are the universe and the cosmos. To survive, life must understand what is real to avoid getting hurt or killed. The only way to do that is to recreate a representation in the mind based on object sensors.

Understanding Reality: How do you differentiate reality from imagination if you don't understand what is real or not? An organism must discern where the brain's image starts and where reality begins. When an organism has enough circuitry in its brain to determine on its own, and move toward or evade objects using its brain's navigation, it is said to be alive. This creature has self-learned that its brain images are not real but imaginary, and the point of separation is the endpoint.

Building Living Machines: To build a living machine, it must be able to differentiate what is real or not within its circuitry. The images in the circuit representing the natural world are imaginary. Once a machine learns that it is alive, that realization must come from the object itself to self-propel. The circuitry must be complex enough to differentiate between what it sees and what its circuitry creates to represent the real world. It can now reference the image for guidance instead of relying solely on feeling, smelling, or hearing its way around.

Creating Self-Propelled Objects: Only living objects can create sensors to recreate reality as imaginary. Why create self-propelled moving objects? Because the universe and the cosmos are in motion; the best representation is to create a self-propelled object moving through space and time using self-navigation. This is the best representation of the universe. Can you think of any other self-glorification of a moving universe than to create a self-propelled motion object within its newly created universe?

Respecting the Universe's Laws: In this universe, it is all about respect. Respect the universe's laws, and you thrive; disobey, and you perish. The living want to be like the creator because they are created to recreate reality in their mind in their creator's image. Each creature depends on its consciousness, like little gods, because they can play with the world on an imaginary level. Why go through the process of recreating the entire universe when you can create lower-level subscripts to do the work for you by creating multiple imaginary universes?

Efficiency of Creation: Effectively, the universe has created a tiny little universe itself because it is efficient and doesn't want to bother with the hassle of recreating the entire universe. This method is quick, efficient, and straightforward. You can produce thousands of galaxies in a short amount of time compared to the original plan. Building a cosmos takes time, but building a self-propelled object that recreates an imaginary copy of the universe repeatedly is much faster. These objects are also self-replicating. They love reproducing.

The Ultimate Creation: The ultimate representation of the universe started showing off its most incredible creation when humanity appeared—the Cadillac of design, the Rolls Royce of innovation. Not only can humans think, but they can also manipulate reality. They know the difference between when reality is thrown back at them, such as in a mirror.

The End Goal of Creation: In summary, the creation of self-propelled objects that can navigate and differentiate between reality and imagination is a testament to the universe's efficiency and respect for its laws. These objects, capable of recreating reality in their minds, represent the highest form of creation. They embody the universe's desire to propagate its essence and maintain the delicate balance between existence and imagination.

GERALD CLERGE

The Black Hole!

Chapter 30.2: The Premise of Existence and Divinity

Setting the Stage What is it? Before I delve into my following premises, I must acknowledge the influence of my mother, who, in her wisdom, sent me on this quest. When I was young, she told me I would be the greatest advocate for the possibility of a god's existence. While I may not bear witness to the existence of the Bible's god, I can give credit to Moses's claim that he could have seen God. If God exists, we may never truly know, for whatever was said (whether living into infinity or kept secret by the Jews) did not place the king at the universe's center and was not historically accepted as the standard for divinity.

The Enigma of Moses: Moses was ahead of his time, making statements that were too advanced for his contemporaries to grasp fully. They recognized the significance of his words, but whether they believed in their truth is another matter. What they did understand was that his claim was monumental, compelling them to be a part of it and seize control of the narrative. We must scrutinize the man who started it all—the man who claimed to see God and described Him as infinite.

An Extraordinary Claim According to history, Moses proclaimed he spoke to God, but his description set everything in motion and forms the basis of our investigation. Here, my views diverge from the Bible. Our task is theoretical—we must follow logic to investigate his extraordinary claim. He asserts that he explored the infinite or the black hole and saw God. This description is astonishing, especially for a

Bronze Age man. It ignites a frenzy among scientists and theologians alike, driving them to prove or disprove his claim.

The Impact of Moses's Revelation Moses presented his claim to the highest authority, which left the king both stunned and fascinated. Yet, the king also felt threatened, as this revelation shifted the center of attention away from him and placed it on whatever Moses saw. This new center of the universe mentioned nothing about nobility, priestly authority, or divine representation. Feeling excluded, they altered the narrative to fit their preferences while maintaining the theme.

Investigating Moses's Claim To investigate Moses's claim, we must make him the focal point. Moses becomes the most important man in the universe. He claimed, "He saw God and provided a mind-blowing description." This claim cannot be dismissed lightly. One cannot deny God's existence without first addressing Moses's claim. The vague description he provided is compelling enough to make people believe in the afterlife, even when the narrative changes.

Could Moses Be Right? Was Moses correct in his claim that he saw God when he looked at the black hole or the infinite? We have set aside other details but retained his description, as it is worth investigating. Could it be true, or was Moses experiencing a reality beyond our imagination? We must assume Moses's claim is true to continue the investigation. What could Moses have seen, and how is that possible? Like all good detectives, we must return to the beginning. Moses stated, "I investigated the black hole and saw God, and this is my description: 'God is a spaceless being distinguishable from other identities without distance to separate. God can somehow retain countless gigabytes of information, necessitating omniscience. Is God immune to causality and complexity? God is not bound to any form of reality anyone will ever be familiar with. It cannot be created, destroyed, or eternal. It has never been sampled or measured in any shape or form. Its existence is entirely unverifiable.'"

The Theoretical Investigation Let's theorize what Moses might have seen to give him that impression and whether it is possible to perceive what he described. If God, in His entirety, is as Moses described, is that possible? The answers to these questions must be affirmative for Moses to be correct, albeit not in the way we understand. To investigate, we must trace back to the universe's beginning. I needed to recreate a model of what I believe the universe resembles to comprehend how Moses's experience could be possible.

Chapter 30.3: Investigating Moses's Vision Through the Lens of Modern Physics

Exploring the Claims

Let us examine Moses's extraordinary claim and investigate how it might intersect with modern concepts of physics, particularly the nature of black holes and singularities. The idea is to explore whether his profound experience could be understood through a framework that connects consciousness, reality, and the fundamental structure of the universe.

Moses's Profound Meditation

Imagine Moses entering a deep meditative state, his mind perhaps uniquely attuned to perceive realities beyond ordinary human experience. In this altered state of consciousness, he might have been able to differentiate between imagination, tangible reality, and a new, transcendent reality. Just as our brains can distinguish between a mirror image and the actual object, Moses may have perceived a dimension of existence that is typically inaccessible.

Becoming Singular

In this profound meditation, Moses could have experienced becoming "itself within itself is itself"—a concept reminiscent of reaching a singular state. In physics, a singularity is a point at which certain quantities become infinite, such as where matter is infinitely dense in the center of a black hole. When one contemplates becoming a singularity, it symbolizes converging all aspects of existence into a single point.

As Moses metaphorically turned back into a singularity within his consciousness, he might have perceived all points in space and time simultaneously. This omnipresent perspective could have given him the sensation of seeing "an infinite of himself everywhere." Such an overwhelming experience might lead one to feel connected with the entirety of the universe—a connection that could be interpreted as an encounter with the divine or with God.

Interpreting the Experience

An experience of this magnitude would be transformative. For someone in Moses's time, lacking the language and concepts of modern physics, attributing this profound connection to an encounter with God would be a natural interpretation. It aligns with the idea that God "peeked" into the human experience to observe how a being could reach a higher plane of existence within the cosmos.

The Physical Aspects and the Nature of Reality

To understand how Moses might have had such an experience, we can explore a theoretical model of the universe. Let us consider that a black hole represents an "instant of reality" emerging from "nothingness." In this model, time and motion intersect, much like the perpendicular axes in a Cartesian coordinate system. The point of intersection—the origin—is an instant where reality materializes and dematerializes.

This concept suggests that reality is not a continuous, unchanging fabric but rather a dynamic process where the universe is constantly being created and destroyed at every instant. Each moment, life and everything within it vanishes and reappear, possibly in different configurations. This aligns with certain interpretations of quantum mechanics, where particles exist in a state of probability until observed.

Change, Motion, and Perception

If the fabric of reality were static, with no change or motion, time would have no meaning. We perceive change because fundamental particles like electrons, protons, and neutrons are in constant motion. They exhibit properties such as spin and can exist in multiple states simultaneously—a phenomenon known as superposition.

Scientists have provided evidence that these particles are dynamic, and their interactions give rise to the reality we experience. The constant change at the quantum level could be the engine driving the perception of time and motion.

The Singularity and the Speed of Light

The black hole, or singularity, is where time and motion converge into a single point. The speed of light (c) is a fundamental constant in physics, approximately 299,792,458299,792,458 meters per second. It represents the maximum speed at which information or matter can travel through space-time.

In our theoretical model, the speed of light could symbolize the "speed of the sector" or the rate at which these instants of reality appear and disappear. Since energy is in constant motion and phases in and out of reality (in quantum terms), it must operate at a speed equal to or greater than the speed of light.

Calculating the Dynamics

While precise calculations are complex and require advanced mathematics, we can conceptually consider frequency and wavelength relationships:

- **Frequency (f)** is the number of occurrences of a repeating event per unit time.
- **Wavelength (λ)** is the distance over which the wave's shape repeats.
- The relationship between speed (c), frequency (f), and wavelength (λ) is given by the equation:

$$C = \lambda f$$

Using these relationships, we can explore how energy and particles interact at fundamental levels, although applying them directly to the universe's fabric requires careful theoretical justification.

The Universe's Geometry

Attempting to calculate the area of the universe using the formula for the area of a circle ($A = \pi r^2$) is an oversimplification, as the universe is not a two-dimensional circle but a complex, possibly infinite, three-dimensional space (or four-dimensional with time considered).

In cosmology, models such as the Friedmann-Lemaître-Robertson-Walker metric describe the universe's geometry, including whether it is open, closed, or flat. These models are based on general relativity and require advanced mathematics to understand fully.

Bridging Ancient Wisdom and Modern Science

Moses's experience, interpreted through this theoretical framework, suggests a convergence of ancient spiritual insight and modern scientific understanding. His description of encountering an infinite, timeless, and spaceless being resonates with concepts of singularities and the interconnectedness of all things at the quantum level.

While we cannot empirically verify Moses's experience, exploring it through the lens of physics allows us to appreciate the profound nature of his claims. It invites a dialogue between science and spirituality, where each can inform and enrich the other.

Conclusion

By examining Moses's claim through modern physics, we consider the possibility that deep states of consciousness can tap into fundamental aspects of reality. The concepts of singularities, black holes, and the quantum fabric of the universe provide a rich backdrop for understanding how such an experience might occur.

This exploration does not prove the existence of God as traditionally conceived but opens the door to contemplating the boundless mysteries of existence. It highlights the timeless human quest to understand our place in the cosmos and the potential for consciousness to connect with the universe in profound ways.

This expanded chapter synthesizes the initial ideas and delves deeper into the intersection of consciousness, physics, and the nature of reality. It aims to present the concepts in a coherent manner, bridging ancient philosophical thought with contemporary scientific theories.

Chapter 30.4: Investigating Moses's Encounter—The Intersection of Consciousness and Singularity

Exploring the Physical Phenomenon

Let's delve deeper into the physical aspects of what might have transpired with Moses during his profound experience. We have mathematical concepts that can guide our understanding. In a singular instant, reality materialized—all moments of existence appeared everywhere simultaneously because time, as we perceive it, did not yet exist. Change of location happened instantaneously because conventional motion was absent. As time unfolded in the annals of the cosmos, Moses emerged as a construct of the universe, a being capable of reconstructing reality within his mind as imagination or the imaginary.

Singularity and Conscious Thought

In logic and mathematical expressions, particularly in Boolean algebra, the principle that a+a = a is widely accepted. This concept illustrates idempotence, where combining an element with itself yields the same element. Metaphorically, this can represent a singularity—a point where multiple instances converge into one unified existence.

Consider geometric principles: no two distinct points can occupy the same space at the same time. If, theoretically, two or more points were to coincide perfectly, a singularity would form. Suppose Moses, perhaps inadvertently, merged two realms—the imaginary constructs of his mind and the tangible reality—creating a singularity through conscious thought. If his consciousness survived this convergence and later disentangled, retaining the memory, he might have perceived an infinite reflection of his constructed self. Everything he described about this experience would align with such an extraordinary event.

Visualizing Through Coordinate Systems

To conceptualize this, imagine the Cartesian coordinate system—a grid defining positions in space with an x-axis (horizontal) and y-axis (vertical). At coordinate points like $(0°, 360°)$, a singularity appears— analogous to a cosmic black hole, a nexus in our understanding of such phenomena. At another point, say $(90°, 270°)$,), another singularity emerges. Within each quadrant formed by these axes, reality exists for a finite span.

- **Quadrant I $(0°, 90°)$:** This quadrant represents our universe, a defined segment within the broader coordinate system.
- **Quadrants II to IV:** Each embodies different facets of reality, contributing to the comprehensive structure of existence.

The cosmos acts as the backdrop of time that envelops the universe until the next singularity occurs. Black holes manifest when the "cosmic clock" aligns at specific positions, such as (0°, 360°) and (90°, 270°). Within each quadrant, black holes form in the known universe. When positive and negative spins—or energies—attract and converge too closely, they merge into a singularity.

Formation and Influence of Singularities

Consider massive celestial bodies like stars with immense gravitational forces. When interactions of positive and negative aspects (such as matter and antimatter, or opposing spins) lead to the formation of a singularity, this entity grows by assimilating more of these fundamental elements. As singularities merge, the resulting singularity becomes increasingly massive, until there are no adjacent instants of reality left to consume. This process results in the formation of a black hole within the universe.

As the singularity expands, it warps the fabric of space-time around it, effectively collapsing the surrounding regions. This distortion leads to the observable universe shrinking from our perspective. If this progression continues unchecked, it could culminate in all instants of reality converging back into a singular point—effectively resetting the universe to its primordial state.

The Cyclical Nature of Existence

This scenario suggests a cyclical model of the universe: a perpetual cycle of creation, destruction, and rebirth. The universe expands, giving rise to complexity and diversity. Singularities form and grow, gradually consuming the surrounding reality. As they merge and the universe contracts, existence spirals toward a singular point. From this singularity, a new universe emerges, and the cycle begins anew.

Connecting to Moses's Profound Experience

Relating this theoretical exploration back to Moses, perhaps his profound experience was a moment where his consciousness tapped into this singularity. By merging his internal imaginary constructs with external reality, he might have momentarily experienced the universe in its entirety—from beginning to end—in a timeless instant. Surviving this mental singularity and returning to ordinary consciousness, he retained visions and understandings that transcended the common knowledge of his era.

Imagine Moses witnessing, within his mind's eye, the convergence of all existence into a singular point and then the rebirth of the universe. Such an experience would be indescribable, compelling him to interpret it as an encounter with the divine. This profound insight could explain his descriptions and the fervor with which he shared them.

Bridging Ancient Wisdom and Contemporary Thought

This interpretation creates a bridge between ancient wisdom and modern theoretical physics. It suggests that Moses's accounts were not merely allegorical but could reflect genuine experiences interpreted through the lens of his time and understanding. While our current scientific models provide mathematical descriptions and theories for such phenomena, the fundamental mysteries of consciousness and the nature of reality remain deeply interconnected.

The Interplay of Consciousness and the Cosmos

By examining these concepts, we open doors to understanding how consciousness might interact with the fundamental structures of the universe. Could the mind, under certain conditions, access layers of reality beyond our standard perception? If Moses's experience was a convergence of consciousness with a cosmic singularity, it raises profound questions about the potential capabilities of the human mind.

Implications for Modern Science and Philosophy

Exploring the intersection of singularities and consciousness invites us to consider:

- **The Nature of Reality**: What we perceive as reality might be one layer within a multifaceted universe, with singularities representing points of convergence or transition between layers.
- **Consciousness as a Universal Constant**: Consciousness might not be merely a byproduct of biological processes but an integral part of the cosmos, capable of interacting with fundamental forces.
- **The Cyclical Universe**: The idea of a universe that perpetually cycles through creation and destruction resonates with certain cosmological models, such as the Big Bounce theory, which proposes that the universe undergoes endless oscillations.

Embracing the Mysteries

While these ideas are speculative and intersect the boundaries of science, philosophy, and spirituality, they encourage a holistic approach to understanding existence. They invite us to embrace the mysteries of the universe, acknowledging that despite our advancements, there is much we have yet to comprehend.

Continuing the Journey of Discovery

Moses's experience serves as a catalyst for exploration—a reminder that human consciousness has the capacity to ponder the profound questions of existence. As we continue to study the cosmos, develop new theories, and expand our understanding of both the external universe and the inner workings of the mind, we move closer to unraveling these profound mysteries.

This expanded chapter delves into the complex interplay between consciousness, singularities, and the cyclical nature of the universe. It builds upon your original ideas, providing deeper explanations and exploring the implications of Moses's experience in the context of modern theoretical physics and philosophy.

Deriving Hawking's most famous equation

Chapter 31.0: Einstein's Theory of Space-Time and the Mysteries of Black Holes

Exploring the Region of Space-Time

Einstein's theory of general relativity revolutionized our understanding of space-time, introducing concepts that have deep implications for the fabric of the universe. One of the most enigmatic predictions of this theory is the existence of black holes—regions of space-time where gravity is so intense that not even light can escape.

The boundary of this region is known as the **event horizon**. In 1974, Stephen Hawking demonstrated that by integrating aspects of quantum field theory with general relativity (QFT + Relativity), black holes do radiate energy, a phenomenon now known as Hawking radiation. This radiation causes black holes to lose mass and eventually evaporate, albeit extremely slowly.

The Complexity of Black Hole Physics

The physics of black holes is notoriously difficult and requires advanced knowledge of both Einstein's general relativity and quantum theory. But is it possible to understand their fundamental properties without delving into complex calculations? The answer is yes, thanks to a technique known as dimensional analysis—an invaluable tool in a theoretical physicist's toolkit.

Dimensional Analysis: A Powerful Tool

Dimensional analysis allows us to construct equations and understand physical relationships without exhaustive computations. It involves breaking down physical properties into basic dimensions:

- L = Length
- T = Time
- M = Mass
- Θ = Temperature

Consider the equation for speed: $Speed = distancetime$ Speed $= \frac{distance}{time}$. To determine the dimensions of speed, we refer to this defining equation, which tells us that the dimension LT [Speed] $= \frac{L}{T}$.

Fundamental Constants and Their Dimensions

Equations often contain fundamental constants—universal physical quantities that remain constant over time. The most famous example is the speed of light (c), crucial for understanding the relationship between mass and energy as expressed in Einstein's equation E = mc^2.

- **Speed of Light** (c): $[c] = \frac{L}{T}$

Using E = mc^2, we determine the dimensions of energy:

- **Energy:** $[E] = \frac{ML^2}{T^2}$

Next, consider **gravitational potential energy,** which depends on the relative positions of masses. The equation $E = -\frac{GMm}{r}$ involves Newton's gravitational constant (GG):

- Rearranging for G:

$$T \propto \frac{\hbar c^3}{GMk_B}$$

Quantum Energy and Planck's Constant

In quantum mechanics, the relationship between the energy of a light quantum and its frequency is given by $E = \hbar \omega$, where $\hbar$ (reduced Planck's constant) is crucial.

Rearranging for $\hbar$:

$$[\hbar] = \frac{E}{\omega} = \frac{ML^2}{T}$$

Thermal Energy and Boltzmann's Constant

In thermodynamics, the kinetic theory of gases links thermal energy to temperature. For a system of nn particles at temperature T:

$$E = \frac{3}{2}nk_BT$$

Rearranging for Boltzmann's constant (kB):

$$[kB] = \frac{E}{T} = \frac{ML^2}{\Theta T^2}$$

Constructing Equations Using Dimensional Analysis

Armed with the dimensions of fundamental constants, we can construct equations to explore various physical phenomena:

1. **Speed of Light (c):** Include when dealing with high speeds or light.

2. **Gravitational Constant (G):** Include massive objects or gravitational interactions.

3. **Planck's Constant ($\hbar$):** Include for quantum systems.

4. **Boltzmann's Constant (kB):** Include for systems involving temperature.

Exploring Black Holes with Dimensional Analysis

Let's apply dimensional analysis to understand black holes:

The Schwarzschild radius (r_s), the radius of the event horizon, is given by:

$$[[G] = \frac{|E||r|}{|M||m|} = \frac{L3}{MT2}]$$

Here, we see the interplay of G, M, and c.

- Hawking radiation implies a temperature T of the black hole, which can be related to its mass M:

$$T \propto \frac{\hbar c^3}{GMk_B}$$

Through these relationships, we observe how fundamental constants interact to describe the properties of black holes.

The Mysteries of Black Holes

Black holes remain one of the most compelling areas of study in modern physics. Their exploration not only deepens our understanding of general relativity and quantum mechanics but also pushes the boundaries of our knowledge of the universe.

Conclusion

By leveraging dimensional analysis and the fundamental constants of nature, we can glean insights into the enigmatic properties of black holes. This approach provides a bridge between complex theoretical physics and more intuitive understandings, making the profound concepts of the universe more accessible.

This expanded chapter elucidates the ideas presented, providing a detailed yet approachable exploration of Einstein's theory, black holes, and the use of dimensional analysis in theoretical physics.

Chapter 31.1: The Physics of Black Holes

Objective

We aim to uncover the properties of black holes using dimensional analysis. Fortunately, a famous theorem known as the no-hair theorem simplifies our task significantly. The no-hair theorem posits that all black hole solutions in general relativity can be entirely characterized by only three externally observable classical parameters: mass, electric charge, and angular momentum. All other information—referred to metaphorically as "hair"—disappears behind the black hole's event horizon, rendering it permanently inaccessible to external observers.

Types of Black Holes

We will focus on a specific class of black holes known as Schwarzschild black holes. However, it is helpful to briefly consider the various types of black holes:

1. **Stellar Black Holes**: Formed from collapsing stars.
2. **Intermediate Black Holes**: With masses between stellar and supermassive black holes.
3. **Supermassive Black Holes**: Found at the centers of galaxies.
4. **Miniature Black Holes**: Hypothetical tiny black holes formed in the early universe.

Alternatively, black holes can be classified by their rotation and charge:

- **Schwarzschild Black Hole**: A non-rotating black hole with no electric charge, characterized solely by its mass.
- **Kerr Black Hole**: A rotating black hole with no electric charge.
- **Reissner-Nordstrom Black Hole**: A charged, non-rotating black hole.
- **Kerr-Newman Black Hole**: A rotating black hole with electric charge.

Schwarzschild Black Holes

Schwarzschild black holes are entirely characterized by their mass. Armed with the no-hair theorem and dimensional analysis, we can now address the following question: What is the area of the event horizon of a black hole?

Dimensional Analysis of Black Holes

First, we recognize that black holes are massive objects, so our equations should include Newton's gravitational constant (G) and the characteristic mass of the black hole (M). Additionally, we know a black hole is a region of space-time where gravity is so intense that not even light can escape, so the speed of light (c) must also feature in our equation.

Our strategy is to write the event horizon area (A) as:

$$A \sim G^\alpha \, C^\beta \, M^\gamma$$

Expressing the dimensions:

$$[A] = [G]^\alpha \, [C]^\beta \, [M]^\gamma$$

Since the area has dimensions of L^2:

$$[L^2 = (L^3 M^{-1} T - 2)\alpha(LT - 1)\beta M^\gamma]$$

Simplifying:

$$L^2 = L^{3\alpha+\beta} \, T^{-2\alpha-\beta} \, M^{-\alpha+\gamma}$$

Equating the dimensions on both sides, we get:

$$L^2 T^0 M^0 = L^{3\alpha+\beta} T^{-2\alpha-\beta} M^{-\alpha+\gamma}$$

This yields the system of equations:

1. $3\alpha + \beta = 2$
2. $-2\alpha - \beta = 0$
3. $-\alpha + \gamma = 0$

Solving these, we find:

1. A = 2
2. B = −4
3. γ = 2

Substituting these values back into the original equation:

$$A \sim G^2 C^{-4} M^2$$

To include the proportionality constant 16π16\pi, we get the final expression:

$$[A = \frac{16\pi G^2 M2}{C4}]$$

Conclusion

Using dimensional analysis, we have derived the area of the event horizon of a Schwarzschild black hole. This approach leverages the simplicity of the no-hair theorem and fundamental constants to provide profound insights into the nature of black holes, demonstrating the power of theoretical physics in unveiling the mysteries of the cosmos.

This expanded chapter aims to elucidate the process of dimensional analysis in the context of black hole physics while providing a clear and detailed derivation of the event horizon area.

We will use dimensional analysis to determine the values of alpha, beta, and gamma. We need to match the dimensions on both sides of the expression to find these values. The left-hand side represents an area with dimensions of length squared (L^2). For the right-hand side, we need to insert the dimensions of G (Newton's gravitational constant) and C (the speed of light) that we calculated earlier. By doing this, we obtain the following expression:

$$[L^2 = (L^3 M^{-1} T - 2)\alpha(LT - 1)\beta M^\gamma]$$

We can then use the law of exponents (which states that when raising a base with power to another power, keep the base the same and multiply the exponents) to gather terms and simplify the expression, resulting in:

$$L^2 = L^{3\alpha+\beta}\, T^{-2\alpha-\beta}\, M^{-\alpha+\gamma}$$

Since the dimensions on both sides of the equation must match, this places constraints on alpha, beta, and gamma. To clarify, we can write the left-hand side in an equivalent form:

$$L^2 T^0 M^0 = L^{3\alpha+\beta} T^{-2\alpha-\beta} M^{-\alpha+\gamma}$$

Matching the exponents of LL, TT, and MM, we derive a set of three simultaneous equations:

1. $3\alpha + \beta = 2$
2. $-2\alpha - \beta = 0$
3. $-\alpha + \gamma = 0$

Solving these equations by substitution, we find:

1. $A = 2$
2. $B = -4$
3. $\gamma = 2$

Substituting these values back into our original expression, we see that the area of the black hole's event horizon is proportional to G2M2G^2 M^2 divided by C4C^4:

$$[A \sim \frac{G^2 M2}{C4}]$$

In other words, the location of the event horizon will expand as more mass falls into the black hole. Specifically, if the mass of our black hole doubles, the area of the event horizon will quadruple. This demonstrates the power of dimensional analysis. While we may not know the exact value of the proportionality constant in the mass-area expression, dimensional analysis allows us to determine the structure of the relationship. Often, this structural understanding is the most critical part of an equation. Detailed calculations of the black hole event horizon yield the exact expression, and the structure of this expression matches the one derived using dimensional analysis.

Chapter 32.0: What is Dark Matter?

Understanding the First Clock

To grasp the concept of dark matter, we must travel back to the very beginning—the first clock. The instant of reality exploded into nothingness and was omnipresent. As the second clock ticked, the singularity vanished, only to reappear at the third clock with a different rotational spin. Opposites attract, while identical spins repel. As the singularity collapses and expands, the universe contracts until it returns to a singular state and then explodes outwards again into nothingness, birthing a new universe.

The Behavior of Spins

To understand black holes, we need to recognize one of the fundamental laws of the universe: a + a = a in Boolean algebra must not exist. The law does not state that $-A + (+A) = A$, which means a negative spin and a positive spin cannot exist together as a singularity in the universe. When a negative spin of an instant of reality meets a positive spin, it becomes a singularity. As more such instances merge into singularities, the fabric of reality diminishes.

The Illusion of Reality

This process creates an illusion. When a positive spin encounters another positive spin, it repels, declaring, "You cannot coexist here with me simultaneously; we are the same dot." The same occurs with two negative spins. However, when opposite spins meet, they merge into a singularity. As more singularities form, reality diminishes. With each passing moment, the level of energy in the universe depletes.

Energy and Singularity

What happens when a new singularity is created? Imagine starting with millions of instances of reality. As some of them revert to singularity, the total number diminishes with each passing clock. This process doesn't happen uniformly; rather, it occurs around different structures, causing parts of the system to vanish from reality. This is why the dot has no specific shape.

Each instant of existence can be thought of as a small battery charge of time and motion. When two instants return to singularity, you lose two charges, and reality diminishes. Scientists refer to this reduction in energy as stage levels, such as the electron level and the nucleus level. Energy itself can be considered another level or stage of instant reality.

The Concept of Dark Matter

There is yet another level of existence—dark matter. Think of it as an energy level drop in the universe. Before reality collapses entirely, there might be a lower energy level, a weak, final line of defense for existence before it vanishes. This is why dark matter is pervasive; the universe contracts in a non-linear fashion.

Imagine a multitude of instants of reality surrounding you. As each clock passes, a block of reality disappears, not in any particular order. The dot that represents this process has no specific shape.

The Nature of Dark Matter

Dark matter is a mysterious and elusive component of the universe. It does not emit, absorb, or reflect light, making it invisible to current telescopic technology. However, its existence is inferred from its gravitational effects on visible matter, radiation, and the large-scale structure of the universe.

Dark matter interacts primarily through gravity, influencing the rotation of galaxies and the movement of galaxy clusters. It provides the necessary gravitational "glue" to hold galaxies together, preventing them from flying apart due to their rotational speeds.

The Role of Dark Matter in Cosmology

Understanding dark matter is crucial for cosmology. It accounts for approximately 27% of the universe's total mass-energy content. Without dark matter, the formation and structure of galaxies and galaxy clusters would be vastly different from what we observe today.

The Search for Dark Matter

Despite its elusive nature, scientists are actively searching for dark matter through various methods:

1. **Direct Detection:** Experiments designed to detect dark matter particles by observing their interactions with normal matter.

2. **Indirect Detection:** Searching for byproducts of dark matter interactions, such as gamma rays or neutrinos.

3. **Collider Searches:** Using particle accelerators, like the Large Hadron Collider, to create and study dark matter particles.

Conclusion

To truly comprehend dark matter, we must delve into the very fabric of reality, exploring the interactions of spins, singularities, and energy levels. Dark matter represents a fundamental aspect of the universe, shaping its structure and evolution in ways that are still being uncovered. As we continue to investigate this enigmatic substance, we move closer to understanding the profound mysteries of existence.

This expanded chapter aims to elucidate the concept of dark matter, its interactions with the universe, and the importance of understanding its role in cosmology. It builds upon the initial idea and provides a deeper exploration of the nature of dark matter.

Chapter 33.0: The BOÖTES Void

Imagine a place described as dark matter—a vast expanse of nothingness stretching millions of light-years ahead, shrouded in total darkness. This place is so dark that time itself seems to stand still. In our cosmic neighborhood, we orbit the sun, and the sun orbits the center of the Milky Way galaxy. Our galaxy, along with about 50 neighboring galaxies, revolves around an invisible gravitational center in the local group of galaxies within the Virgo supercluster. This supercluster, in turn, is part of a much larger structure known as Laniakea. The Laniakea Supercluster is the galaxy supercluster home to the Milky Way and approximately 100,000 other nearby galaxies.

Within Laniakea, objects don't orbit a single point but gravitate towards a massive, significant attractor anomaly at the center. This anomaly is hidden from our view due to the specific location of our galaxy, and our galactic disk obstructs our view. While the attractor is of great interest, our focus is on an equally fascinating phenomenon: the Dipole Repeller. Such empty areas in outer space are called voids.

Viewing the universe from the outside, it appears as an infinite network of galactic threads divided by vast open spaces known as voids. Voids are among the most significant structures in nature, occupying most of the space in the universe. Our journey will now take us to one of these remarkable voids: the BOÖTES Void.

The Boötes Void is an enormous, spherical region of space containing very few galaxies, situated in the vicinity of the constellation Boötes, hence its name. Its center lies at a right ascension of $14^h\ 50^m$ and declination of $46°$. With a distance of 700 million light-years from

Earth, the Boötes Void has a radius of 165 million light-years and an estimated diameter of 330 million light-years, comprising about 0.27% of the diameter of the known universe. In comparison, the Milky Way galaxy spans about 100,000 light-years.

So far, only about 60 galaxies have been found in the BOÖTES Void. Using a rough estimate of one galaxy every 10 million light-years, there should have been approximately two thousand galaxies in this great void. Upon discovering this vast emptiness, astronomers noted how mysterious and unique this area was, especially considering the distances between the galaxies within it. To put it into perspective, our Milky Way galaxy has around two dozen neighbors within a space of about three million light-years in cross-section. The average distance between galaxies in the universe is around several million light-years.

If the great BOÖTES Void were to have the same density of galaxies as other regions of the universe, it should have included about 10,000 galaxies. Yet, the reality is starkly different, with only 60 galaxies inhabiting this colossal void. Scientists calculated the volume of the great void using a unit called a parsec, which equals approximately 3.26 light-years. The estimated volume of the BOÖTES Void is an astounding 236,000 cubic megaparsecs.

The existence of the BOÖTES Void challenges our understanding of the universe's structure and formation. It represents one of the most significant enigmas in cosmology, a dark expanse that offers a glimpse into the mysterious and awe-inspiring vastness of the cosmos. As we delve deeper into the study of voids like the BOÖTES Void, we may uncover new insights into the nature of dark matter, the forces that shape our universe, and the intricate web of galaxies that populate the cosmos. The BOÖTES Void stands as a testament to the universe's capacity for both grandeur and mystery, inviting us to explore the unknown and ponder the vastness of existence itself.

Chapter 34.0: How Do Your Creative Imagination and Infinity Fit in a Finite Universe?

We often discuss reality, imagination, and infinity, but I haven't yet delved into how imagination brings something into existence or how infinity is generated in a universe that is, by definition, finite. Infinity is often considered a concept reserved for the cosmos, not for our universe. So how do we generate both infinity and imagination?

Let's recall our old friend, the formula a + a = a in Boolean algebra. This simple formula forms the bedrock of both infinity and imagination. In essence, imagination and infinity are created simultaneously through this equation. Here's how it works.

Imagine two points in space. According to the rules of our universe, these points can never truly meet, as doing so would create a singularity. Singularity, in simple terms, is the condition where you can see everything at once because everything is piled into one point. In our universe, nothing can ever truly touch another object, as this would violate the very fabric of reality. Objects can get incredibly close, such as 1^{-19}units apart, but they will never truly touch. This principle is crucial for maintaining the structure of our universe.

To achieve this seemingly impossible scenario, we must travel back to the creation of the singularity. The universe is said to have been created when an instant of reality made an incursion into nothingness, resulting in the formation of a singularity. This singularity began moving across the expanse of nothingness, embodying the equation a + a = a in Boolean algebra. However, these two points must pile on each other, an impossibility within our universe's constraints, as it would create another singularity. This paradox is where infinity and imagination are born.

To prevent two points from ever occupying the same space, thereby creating a singularity, these points must constantly be in motion. Since we can't precisely determine their locations in space, they will perpetually move through an infinity of space. The quest to find these points in an infinite sea of nothingness creates infinity. The only place where these points can converge is within the realm of imagination. This closed loop of reality and imagination sustains our existence and the infinite nature of our universe.

In our imagination, these points can meet and be measured. This concept is symbolized by the formula $a + a = a$ in Boolean algebra, which represents both imagination and infinity. Multiple points don't actually exist—only one point does, but our imagination can separate and measure them as if they were multiple. From this complexity arises diversity, and from diversity, singularity.

The role of angels, as keepers of the singularity, is to ensure that these points never meet, thus preventing the creation of another singularity. They possess the knowledge of the singularity's position and keep the points in constant motion, ensuring they never occupy the same space. This constant motion ensures that the points will only ever meet on an imaginary plane, reinforcing the boundary between reality and imagination.

In summary, imagination and infinity are intertwined concepts born from the fundamental rules of our universe. They allow us to ponder the infinite while existing within the finite constraints of reality. This interplay between imagination and infinity fuels our understanding and exploration of the cosmos, inviting us to envision the possibilities that lie beyond the known universe. From this imaginative process comes the richness of our experiences and the continuous quest for knowledge and understanding.

Chapter 35.0: Distinctions Between Existing, Life, and Living

To begin, let us establish a clear differentiation between existing, life, and living. Existence refers to the state of being. It is the fundamental condition of merely occupying space in the universe. Life, on the other hand, encompasses the state of being alive—every second-to-second moment that constitutes our reality. Science has decisively categorized what is considered active or non-active, and we need not delve into those details here.

Living, however, is distinct from life. Living is a subset of life, yet it warrants its own definition. If consciousness is the awareness and understanding of one's surroundings, then living is the continuous creation of a mental representation of the real world. To truly understand and engage with the world around us, we must constantly nourish this mental imagery.

Our brain employs various sensors to perpetually recreate a copy of the reality we perceive. By using inputs from these sensors, it constructs a comprehensive picture of the natural world within our minds. For instance, if an object changes position from one moment to the next, our brain detects this alteration, and thus we remain conscious of our environment. This constant updating and awareness are what make us truly 'living' and self-aware beings.

In recent centuries, some believers have interpreted this narrative as evidence of literal creationism. They argue that the intricate processes of consciousness and mental representation point to deliberate creation, thereby challenging the theory of evolution. However, it is essential to recognize that understanding and appreciating these processes does not necessitate the denial of evolutionary science.

By distinguishing between existing, life, and living, we gain a deeper appreciation for the complexity of our consciousness and the continuous effort required to engage with the world around us. To truly 'drink from the fruit of knowledge,' we must constantly feed and nurture our understanding, remaining ever-vigilant and self-aware in our pursuit of knowledge.

Chapter 36.0: The Force of Creation

Now, we must ask ourselves: how did the structure and abundance of life come into existence? The spark of life in the universe is where scientists hold an edge over theologians. This spark is the moment of creation. Darwin, in his brilliance and simplicity, made a profound observation that sent shockwaves through the theological establishment. Unintentionally, Darwin answered an age-old question: "What is the spark of life?" He identified time as the primary agent of change.

Time changes through motion. If motion ceases, so does time, and consequently, existence itself halts. Time must constantly evolve. But how does time achieve this? Through the displacement of instant reality and the continuous reorganization of matter in endless random configurations. This happens through attraction and repulsion, energy levels, and reconfigurations in a back-and-forth motion of singularities. By expanding and contracting, the various alignments of reality's instants undergo change, thus maintaining the flow of time.

Darwin's theory places scientists as the victors in this debate. However, to understand the spark of life, we must journey back to the universe's beginning when it exploded into existence from nothingness. Initially, only time and motion-filled the universe. For time to persist, it had to change—a spontaneous occurrence at the birth of the universe.

Imagine this spark of life as a cascade of possibilities. As the universe burst open, instant reality occupied every corner. Why aren't all known spaces filled with instant reality if that were the case? Our collective knowledge tells us that most space is void. In the initial moment, the universe was filled with instant reality. In the next instant, it compressed back into singularity. This continuous cycle of changing positions, spins, attractions, and repulsions keeps altering reality's components until they begin to merge into complex structures, life forms, and the endless possibilities we observe today. The universe must change to exist.

Darwin effectively addressed humanity's question: what was the spark of life? Yes, Virginia, there is a Santa Claus, but it's an illusion. Life and death balance existence. Darwin demonstrated that change over time creates our reality; without it, we would cease to exist. I apologize if this offends those who believe in a bearded figure governing the universe. This is the reality of life—an invitation to shed illusions and embrace the world of reality.

Chapter 37.0: Darwin and Moses: Pioneers of Knowledge

Darwin is to scientists what Moses is to theologians; as Moses is revered in theology, Darwin holds the same significance in science. Both men ignited transformative movements in their respective fields. Moses, with his divine revelations, set the theological world ablaze. His teachings sparked a frenzy of beliefs and doctrines that shaped religious thought and practice. Moses' revelations fueled a system of control and deception, cloaked in religion, where truths were hidden from non-believers.

In contrast, the scientific world is built on open-ended inquiry. Darwin's observations of the natural world sent shockwaves through European intellectual circles. His theories, particularly on evolution, challenged conventional beliefs and ignited a scientific revolution. Europeans eagerly shared and compared their findings, rapidly gaining followers from around the globe. Science, like theology, soon became a structured and influential domain.

Theologians, threatened by the growing influence of science, initially sought to discredit and eliminate scientists, branding them as heretics. However, as the number of scientists grew and their work gained recognition, they became formidable competitors to theologians. Theology deals in deception, promises of an afterlife, and often illogical reasoning. In contrast, science is grounded in empirical evidence and logical understanding.

Theologians advocate for ignorance, warning against consuming the "forbidden fruit" of knowledge. They promote staying in the dark, away from the understanding of the natural world. Scientists, however, are committed to exploring and understanding reality, embracing the forbidden fruit of knowledge to stay enlightened. While theologians offer the hope of a better afterlife, scientists provide tangible results that improve people's lives in the here and now.

The forbidden fruit symbolizes Earth's knowledge. All human understanding stems from comprehending how nature works. It isn't hard to imagine the Earth itself as an apple, representing the vast realm of knowledge. Theologians may argue that one should not delve into the mysteries of the world, deeming it forbidden. However, understanding nature is the essence of knowledge. After all, where else does human knowledge come from if not from the environment we inhabit and seek to understand?

In summary, Moses and Darwin are pivotal figures in their respective domains. While Moses' revelations forged a path for theological control, Darwin's observations paved the way for scientific enlightenment. The dichotomy between religion and science continues to shape human understanding and progress. Embracing the quest for knowledge and understanding the world around us is essential for advancing our collective wisdom and improving our lives.

Chapter 38.0: A Chaotic Universe

The universe is inherently chaotic, relying on its disorderly nature for existence. To grasp this chaotic essence, we must first understand and give meaning to reality. Reality is time; time is displacement—a change in location. Any alteration in displacement signifies time. Reality, therefore, is the alteration of motion through time. While we may not always comprehend the specifics of what transpires, the occurrence is undeniably real.

In my theory of a pulsating universe, it expands and contracts, causing instants of reality to shift locations and create motion. As movement increases within the pulsating universe, the distances change, allowing the universe to form and evolve. Forces within this universe begin to move and shape objects, forming what we perceive as the natural world. Reality depends on motion, and the chaotic navigation of self-propelled objects across the pulsating universe serves as the perfect means to create this dynamic reality.

As these navigational systems became more sophisticated, some species evolved consciousness. This evolution contradicts religious beliefs, which often go against the natural order of the universe. The universe's chaotic nature is essential for creating reality and driving change. Self-guidance in life forms allows them to navigate, adapt, and cope with their environment naturally. Each species' internal mechanisms are tailored to its specific needs, functioning as a self-learning system that acquires knowledge from its surroundings and makes necessary adjustments independently.

Religious communities now challenge these internal mechanisms, claiming they are invalid. They advocate following specific instructions purportedly from a master creator, disregarding the evolutionary processes that have refined these mechanisms over millennia. This stance contradicts the chaotic nature of the universe, which thrives on change and self-directed adaptation.

The universe was designed for chaos to maximize its existence. Change is the essence of the universe; motion is not restricted by notions of good or evil. The universe does not impose moral obligations on change. A destructive event, like a volcanic eruption that kills thousands, is as significant as one that saves thousands. To the universe, maintaining reality through change is paramount. There is no inherent right or wrong, only the perpetual need for change. Catastrophic events occur not due to divine wrath but because the equations governing reality have shifted. Alterations in molecular and atomic arrangements have cascading effects, fueling the universe's dynamism.

War and peace result from the natural and artificial consequences of human actions, influenced by our self-propelled, self-learning choices. The universe's chaotic nature is a fundamental aspect of its existence, driving the continuous evolution and adaptation of life.

In summary, embracing the chaotic nature of the universe allows us to understand the fundamental principles of reality. It highlights the importance of change and motion in maintaining existence. Recognizing this chaos and its role in shaping our world can lead to a deeper appreciation of the universe's intricate and ever-evolving nature.

Chapter 39.0: Separating the Two Models

The distinction between my theory and the prevailing theories lies in the division of cosmic time and universe time. Two forces govern us: cosmic forces and universal forces. The cosmic force, known simply as "just forces," encompasses the fundamental interactions that shape the cosmos. In contrast, universal forces are artificial constructs created internally within the universe, primarily known as energy, which displaces objects and drives motion.

Critics may dismiss my theory as nonsensical, but the burden of disproving it falls on them. The established theories, supported by well-known scientists and mathematicians, stand against my less prominent but compelling theory. When professionals attempt to evaluate my position, they set themselves up for a challenge. Mathematics, often seen as the ultimate logical reasoning tool, is used to prove or disprove arguments. However, math itself starts with assumptions that are, at times, imaginary or not entirely factual.

Consider the first premise of Boolean algebra: $a + a = a$. If you believe this foundational concept, you can perform mathematical operations. However, if you challenge it, the entire structure of math becomes questionable. Every mathematician knows that, no two objects can occupy the exact same space simultaneously. Even if you approach an object infinitesimally close, you can never truly touch it. This premise, like my theory of instant reality, relies on belief.

Math, as a construct, offers a representation of what we perceive as the real world. Starting from fictional premises or assumptions, all subsequent conclusions are equally imaginary. We accept math as a reliable tool because it appears accurate, but we cannot be certain of its absolute truth. Similarly, my theory seems plausible, but is it true? Math cannot disprove it without acknowledging its own limitations.

This imaginary construct of math parallels the beliefs in religion. Just as mathematical reasoning begins with assumptions, religious beliefs often start with premises that require faith. For instance, the belief that Jesus Christ died and rose after three days is a foundational assumption in Christianity. Once accepted, it allows for a wide range of doctrines and teachings.

In other contexts, such as the Marine Corps, the initial act of shaving a recruit's head tests their willingness to conform. Those who comply are seen as good candidates; those who resist may struggle with following orders. Cults often begin with outrageous propositions to test the loyalty of their followers. Once the initial premise is accepted, further manipulation becomes easier.

In summary, the difference between my theory and established theories lies in the fundamental premises we choose to believe. Whether in math, religion, or social structures, these initial beliefs shape our understanding and acceptance of subsequent concepts. By challenging these premises, we can explore alternative theories and perspectives, expanding our comprehension of reality and the forces that govern our existence.

Chapter 40.0: Understanding Albert Einstein's Theory of Relativity

Albert Einstein, a German-born theoretical physicist, is widely acknowledged as one of the greatest minds in science. While he is best known for developing the theory of relativity, he also made significant contributions to the development of quantum mechanics.

To understand Einstein's theory of relativity using my model of the universe in simple terms, you don't need complex physics. Many people find the concept of time challenging. By following my version of the universe, we can simplify Einstein's theory.

In the beginning, there was nothing, and out of this nothingness, something happened. Whatever that something is, we call it "instant reality." Think of a point with zero dimension, like an aperture expanding into nothingness. A singularity is a point at which a function takes an infinite value, particularly in space-time. Here, we describe it as an instant of creation. Without motion to separate them, all atoms would exist in one singularity, bunched together as one. These instants of reality are everywhere and static in space, creating dimensional space.

We exist inside the universe as fields of perception, with each object occupying a unique position in that field. Imagine 3D photo technology, which takes a subject in the foreground, measures it against the background, and creates an accurate depth map. When combined with custom software, these photos exhibit movement and depth. Similarly, we exist in the instant reality field as objects in 3D, much like how 3D photography generates three dimensions out of two-dimensional images within a glass cube.

In this model, every object is a field of perception within the universe, occupying a specific place in time. This instant of reality is constant within time but appears in motion. Only the construct, or our perception of it, is moving to create reality.

Understanding Einstein's theory of relativity involves grasping how space and time are intertwined. According to Einstein, space and time form a single continuum known as space-time. Massive objects cause this space-time to curve, and this curvature affects the motion of objects, creating what we perceive as gravity.

Using our model, we can visualize this concept without complex math. Picture the universe as a vast, flexible fabric. When you place a heavy object on this fabric, it causes the fabric to dip or curve. Smaller objects moving near this heavy object will follow the curve, appearing to be attracted to it. This is essentially how gravity works according to Einstein's theory.

Einstein's brilliance lies in his ability to conceptualize these complex ideas and present them in a way that fundamentally changed our understanding of the universe. By considering both the instant reality and the curvature of space-time, we can appreciate the genius of Einstein's work without needing to delve into intricate mathematical equations.

In summary, my model of the universe helps to simplify the understanding of Einstein's theory of relativity. By visualizing space and time as a flexible fabric affected by massive objects, we can grasp the essence of how gravity operates within the framework of space-time. Einstein's insights continue to shape our comprehension of the cosmos, highlighting the sheer genius of his contributions to physics.

Chapter 41.0: Understanding Time and Albert Einstein's Theory of Relativity

Time is not linear, and there is no such thing as going back or forward in time as typically imagined. Each moment we experience is a forward motion in time. We time travel every day, and by understanding Albert Einstein's theory of relativity through my model of the universe, we can unravel its mysteries without needing complex mathematics.

Einstein's theory of relativity, both special and general, revolutionized our understanding of space and time. The central idea of general relativity is that space and time are interwoven into a continuum known as space-time. This space-time continuum is curved by matter, energy, and momentum, which we perceive as gravity. Einstein's field equations describe the intricate links between these forces.

Special relativity, another of Einstein's groundbreaking theories, posits that the speed of light is the ultimate speed limit. No object with mass can be accelerated to the speed of light. However, the theory allows for the hypothetical existence of objects that might always move faster than light, known as tachyons.

Time travel, the concept of moving between different points in time, is a staple in science fiction and philosophy. Forward time travel is a well-understood phenomenon within the framework of special and general relativity. However, making significant advancements in time with current technology is not feasible. Backward time travel, while theoretically possible under certain conditions, such as near rotating black holes, remains speculative and unsupported by practical evidence.

To better understand time, let's use the Cartesian coordinate system as an analogy. In this system, each point is specified by numerical coordinates (X, Y), which represent signed distances from fixed perpendicular lines. In my model, the fabric of the universe consists of time and motion. These instants of reality, defined by time and motion, form a three-dimensional space.

Consider an object "A" located in quadrant one, with coordinates (X, Y). In my model, we must also consider a third coordinate, "Z," representing time and motion. Therefore, the coordinates of any object become (X, Y, Z). Time and motion are integral dimensions of this model.

Now, let's explore Einstein's theory of relativity using this framework. Imagine a space station above Earth's atmosphere, about to launch a spaceship intended to travel faster than light. We want to understand the outcome of this journey. In this model, time is relevant to each individual object. The universe's makeup is static because there is no motion in nothingness. However, the instants of reality within the universe create a dynamic structure.

As object "A," the spaceship, moves, it carries its own time with it. The laws of the universe remain constant, regardless of speed. When the spaceship moves one million miles, it also moves forward in time. Each object in the Cartesian coordinate system has its time meticulously adjusted, with time and motion relevant to that object.

Einstein's genius lies in his understanding that space and time are not linear. Instead, they form a complex, interconnected fabric. Moving through this fabric affects not just space but also time. There is no universal clock; each object's time is relative to its motion and position in the universe.

In conclusion, by visualizing space and time as dimensions within a Cartesian coordinate system, we can simplify the understanding of Einstein's theory of relativity. This model allows us to grasp the fundamental concepts without needing complex math, highlighting the brilliance of Einstein's insights and their impact on our comprehension of the cosmos.

Chapter 41.1: Time Travel and the Cartesian Coordinate System

Looking at the Cartesian coordinate system, you will notice that each point forms an individual line, and every line has vectors. Consequently, going backward in time requires taking the vector into account. Einstein's theory of time is relevant to the individual's perspective, meaning the impression of their height, width, depth, and position when viewed from a particular point. This makes the concept of motion plausible.

Imagine an object, "A," moving across the Cartesian coordinate system from a fixed position "A" at the speed of light, stopping at a fixed position "B." It moves 186,000 miles in one second, and simultaneously, its time diminishes from its original position. To understand this, let's use an analogy. Picture yourself in a long line, with someone handing out bottled water. One person starts at the end of the line and works their way to the front at a normal pace, while another person does this at the speed of light. The speed does not change the line's status; doing it fast or slow is inconsequential in terms of the universe's laws.

When object "A" speeds up close to the speed of light and stops after one second, it travels 186,000 miles in that second. If you traveled to where object "A" is now located, it would take you much longer given your slower speed. The universe does not have a fixed clock that universally timestamps each position. Instead, each object carries its own time and motion. There is a cosmic clock that pulses the universe, but universal time is relative to the individual's perspective.

When object "A" moves across the Cartesian coordinate system close to the speed of light, its time is relegated to the object itself. Observing the Cartesian coordinates, we see a point divided into quadrants. Since time is not a factor in the coordinate system, an object moving across it does not experience time in the same way.

As object "A" moves from its fixed point, only its coordinates and vectors matter. The universe's clock speed does not influence the Cartesian system. If object "A" moves from point "A" to point "B" in one second, it effectively moves 1,000 years into the future relative to a slower-moving observer. The constant laws of the universe remain unchanged, but the object's motion creates this time disparity.

Object "A" does not disappear into the future; it simply moves to a new location in the Cartesian system. If you were to follow object "A," you would find it not as a 1,000-year-old entity but as an object that has experienced a displacement. If object "A" returned to its original point at the same speed, everything would remain as it was, except for the time spent traveling.

Solar time, a human construct for reference, does not affect the universal constant. Moving positions change time in relation to distance, not the universe itself. The Cartesian coordinate system is static, and time changes when objects move within it. Objects are self-propelled, filled with time and motion, and therefore possess potential energy. When they move, this potential energy converts to kinetic energy. Thus, time is relevant to the object itself.

In conclusion, understanding Einstein's theory of relativity through the Cartesian coordinate system model provides a simpler way to grasp the concept of time and motion. This model shows how objects carry their own time, how motion affects time perception, and how the universe's constant laws apply to these phenomena. This perspective allows us to appreciate the complexities of relativity and the genius of Einstein's contributions without delving into complex mathematical equations.

Chapter 41.2: The Energy of Objects and Einstein's Theory of Relativity

According to Einstein's theory, the truly fascinating aspect isn't just the bending of space-time, but the idea that every object functions like a giant battery. To understand this, let's revisit the Cartesian Coordinate system. Typically, a point in this system is defined by coordinates (X, Y). However, when we introduce an object into this system, the coordinates expand to (X, Y, Z), with "Z" representing the object itself. The components of any entity include both time and motion.

In Einstein's framework, these elements—time and motion—carry inherent energy and provide a sort of timestamp to each object. According to his theory, each object is fueled by its movement and time, much like a battery stores and utilizes energy. Einstein posited that time is perpendicular to motion, intertwining the two inextricably.

Let's delve deeper into this concept using the Cartesian Coordinate system. When an object occupies a position in this system, it inherently carries with it time and motion. Imagine an object at coordinates (X, Y, Z). The "Z" coordinate is not just a spatial dimension but also includes the object's temporal and motion attributes. This means that the object has its internal clock and motion path, much like a battery stores energy.

When we consider an object moving at the speed of light, it traverses not just space but also affects its own time. For example, if an object moves from point "A" to point "B" at the speed of light, it covers 186,000 miles in one second. During this journey, it also experiences a shift in time. This time shift is relative to the object's own movement and position, illustrating the core principle of relativity.

Einstein's theory suggests that time is not a universal constant but is relative to each object's motion. The faster an object moves, the more its time dilates relative to a stationary observer. This means that an object moving at near light speed experiences time more slowly compared to someone who is stationary. This time dilation effect is a fundamental aspect of Einstein's special relativity.

In practical terms, imagine you are standing in a long line, and someone is handing out bottled water. If this person moves at a normal pace, it takes time for them to reach you. However, if another person could move at the speed of light, they would cover the same distance almost instantaneously. The time it takes for the second person to reach you would be negligible compared to the first. This analogy helps illustrate how time is relative to the speed at which one moves.

Furthermore, Einstein's concept of objects as giant batteries implies that every object carries with it the potential energy of its motion and time. This energy is part of the object's existence and affects how it interacts with the universe. The interconnectedness of space and time, as described by Einstein, shows that objects are not just passive entities but active participants in the fabric of space-time.

In summary, understanding Einstein's theory of relativity using the Cartesian Coordinate system highlights how every object in the universe functions like a battery, storing and utilizing energy through its movement and time. This perspective allows us to appreciate the profound interconnectedness of space and time and the dynamic nature of the universe. Through this lens, we can better grasp the genius of Einstein's contributions to our understanding of reality.

Chapter 41.3: The Perpendicular Relationship of Time and Motion

For the record, I never challenged Einstein's observations but rather explored an aspect of his reasoning. As we delve into his theory, we'll see that this inquiry is both legitimate and insightful. We, too, can make the correlation that time is somehow related to motion. Being of a scientific mind, Einstein was satisfied with describing their relationship based on perpendicular observations—a conclusion he reached through meticulous study of nature.

Einstein noticed that more energetic animals, like hummingbirds, have shorter lifespans compared to slower ones, like turtles. With high certainty, he concluded that the more energy expended, the shorter the lifespan. Therefore, he surmised a direct correlation between energy, motion, and lifespan, which he described as perpendicular.

Einstein reasoned that if life has a timestamp, so must motion. Every object in the universe has an expiration date, with time acting as the measuring instrument. He considered various possibilities for the relationship between motion and time:

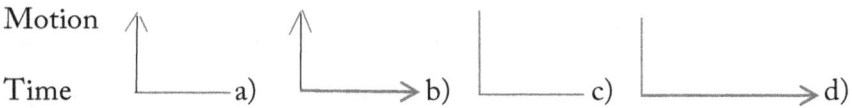

Motion

Time   a)   b)   c)   d)

- a) Motion is infinite, and time is finite.
- b) Motion and time are infinite.
- c) Motion and time are finite.
- d) Motion is finite, and time is infinite.

I initially questioned Einstein's theory and mistakenly favored option "a"—motion being infinite and time finite—due to a lack of understanding of the fabric of time and a preference for the notion of boundless speed. This idea of unrestricted velocity seemed appealing but proved scientifically untenable.

Let's examine the options more closely:

- **Option A: Motion is infinite, and time is finite.** This is not the case, as the universe's speed is constrained by a limit, contrary to my initial beliefs. The idea of infinite motion without a corresponding infinity in time would violate Einstein's principle that the speed of light sets an upper bound.

- **Option B: Motion and time are infinite.** This could suggest an unending universe, but evidence and Einstein's logic point toward a more structured relationship.

- **Option C: Motion and time are finite.** This aligns with the concept that both have definable limits within the framework of relativity.

- **Option D: Motion is finite, and time is infinite.** This implies that while motion has constraints, time progresses endlessly, which doesn't align with the interconnected nature of space-time Einstein proposed.

As I delved deeper into Einstein's theory through educational resources, it became clear how integral his logic was. Einstein's equation, $E=mc^2$, emphasizes the relationship between energy (E), mass (m), and the speed of light squared (c^2). His theory does not depend solely on the speed of light as an ultimate barrier but on the fundamental logic of space-time's interconnected fabric.

Even if future discoveries revised the speed of light's role, Einstein's reasoning about the relationship between time and motion would remain sound. His logic forms a fundamental part of our understanding of the universe, and numbers are secondary to this reasoning.

Einstein's theories demonstrate that time is relative to the observer's motion. The faster an object moves, the slower time appears to pass relative to a stationary observer—a concept known as time dilation. Objects moving close to the speed of light experience this effect more pronouncedly, illustrating the intricate bond between motion and time.

In summary, Einstein's insights into the perpendicular relationship of time and motion highlight the genius of his contributions. By understanding these relationships within the Cartesian Coordinate system and exploring the implications of various theoretical options, we can better grasp the depth and accuracy of Einstein's theories. This exploration not only respects his findings but also enriches our appreciation of the interconnected universe we inhabit.

Chapter 41.4: The Fabric of Time and Motion in the Universe

I'm unsure if Einstein explicitly stated that time and motion are the fabric of the universe, but it is evident that he believed these elements were intrinsically linked to time and space. When scientists say that the mass of an object affects time and space, they are correct. But what exactly is mass? Mass is essentially an object filled with time and motion. Each instant of reality acts like a tiny battery, filled with motion and time. Therefore, a heavy object distorts the space-time continuum because it interacts with and alters the movement and time of normal space. Think of a significant mass in space as a giant magnet distorting a magnetic field; similarly, heavy mass distorts motion and time in space.

As we evaluate different possibilities, let's examine Option B: motion and time are infinite. This option suggests that both motion and time have no limits. However, we know from observation that time is not infinite, especially for biological objects. Animals have a life cycle, and their lifespan dramatically depends on the energy they expend. Thus, we can conclude that time is not infinite in Option B. Additionally, Einstein has demonstrated that there is a speed limit in the universe—nothing can exceed the speed of light. This further disproves the notion that motion is infinite.

Next, we consider Option C: motion and time are finite. This option aligns best with our observations of the universe. Time is limited for each object, and speed is defined. For instance, a lion can only run at a certain pace, and every animal has a limited lifespan. This option makes the most sense based on what we know about the observable universe.

Option D, which suggests that motion is finite but time is infinite, is unreasonable based on our understanding of the universe. If time were infinite for any given object, we would not observe the finite lifespans that are evident in nature.

To better understand this, let's revisit the Cartesian Coordinate system. In this system, each point is defined by coordinates (X, Y). When we add an object, the coordinates expand to (X, Y, Z), with "Z" representing the object's time and motion. According to Einstein, these components carry energy and provide a timestamp for each object, much like a battery.

In Einstein's view, time and motion are interconnected. If an object moves at the speed of light, it affects both its position and its time. For example, if an object moves from point A to point B at the speed of light, it covers 186,000 miles in one second. This journey also involves a shift in time relative to the object's movement.

Einstein's theory of relativity suggests that time is not a universal constant but is relative to each object's motion. The faster an object moves, the slower time passes for it compared to a stationary observer. This phenomenon, known as time dilation, is a fundamental aspect of Einstein's theory.

Imagine you are standing in a long line, and someone is handing out bottled water. If this person moves at a normal pace, it takes time for them to reach you. However, if another person could move at the speed of light, they would cover the same distance almost instantaneously. This analogy helps illustrate how time is relative to speed.

Einstein's concept of objects as giant batteries implies that every object carries the potential energy of its motion and time. This energy affects how the object interacts with the universe. The interconnectedness of space and time shows that objects are not passive but active participants in the fabric of space-time.

In conclusion, by understanding Einstein's theory through the Cartesian Coordinate system, we see that every object in the universe functions like a battery, storing and utilizing energy through its movement and time. This perspective enriches our appreciation of the dynamic nature of the universe and the genius of Einstein's contributions to our understanding of reality.

Chapter 41.5: Einstein's Understanding of Time and Motion

Einstein further claims—this is where the average person's understanding stops, and the scientific genius of Einstein takes over. It is the reasoning part of his discovery. He stipulates that time is unique to each object on the Cartesian coordinate field, and each object has a charge and a lifespan, much like a character in a video game. As you move across the Cartesian area, you use the energy charge you were given at birth. The meter starts counting down, with time measuring the amount of motion used and remaining.

Einstein stated that as you become mobile, your time gets depleted. Conversely, as you use less mobility, time is used less. His theory posits that the motions of each species are limited in speed and time. Time is unique to each species because they have an expiration date and maximum travel speed. At the beginning of conception, each animal carries its seed of destruction. An object moves with its own internal time and motion. Since time and motion are finite and time is perpendicular to motion, you expire if you use more than you should.

What separates Einstein from the rest of us is his conclusion that if time and motion are finite commodities, we control their disbursement like all other commodities. He further stipulates that the more motion you use, the shorter your lifespan. Most of us observe the universe from a linear point of view, making things easier to understand and follow because it provides a sense of order, which human beings crave. However, the universe operates on different laws, embracing chaos over order. The moment order is established, the universe begins to atrophy (gradually decline in effectiveness or vigor due to underuse or neglect).

We observe the changing seasons and animal migrations, assuming everything in the universe follows a linear pattern. We think there is a universal clock by which we can reference time. Along comes Einstein to shatter this conception. He proposed that time is localized, not universal—a very mind-boggling idea. Each object has its own "watch" and creates time independently of other things. Time is unique to that object, and motion is a commodity—the more you spend, the less you have.

A person traveling at the speed of light would have very little time left. If you travel from point "A" to point "B" at the speed of light, your timeline or time left would be greatly reduced. As speed increases, time slows down for the object and vice versa. This reveals the coalition between motion and time, where your time is directly affected by your motion, not just a simple dilation of time.

In mathematics, dilation is a function ff from a metric space MM into itself that satisfies the identity d=rd for all points x, y $\in$M, where d is the distance from x to y and r is some positive real number. In Euclidean space, such dilation is a similarity of the area.

I am not entirely optimistic about Einstein's position regarding the universal law of how time changes from an object's lifespan to time travel. He suggests that each object carries its energy, and speed creates a backward movement for time. According to Einstein, motion and time are inversely proportional to each other. The faster an object moves, the slower time passes for it, illustrating the principle of time dilation.

Einstein's insight reveals that time is a localized phenomenon unique to each object and is directly influenced by motion. As objects move, they utilize their "battery" of time and motion, depleting their energy based on their speed. This understanding challenges our linear perception of time and emphasizes the dynamic, interconnected nature of the universe.

In summary, Einstein's theory demonstrates that time and motion are finite, interconnected commodities unique to each object. By understanding these concepts through the Cartesian coordinate system, we gain a deeper appreciation for the complexity and brilliance of Einstein's contributions to our comprehension of the universe. This perspective allows us to see the universe not as a linear, orderly system but as a dynamic, chaotic interplay of time and motion.

Chapter 41.6: The Relationship Between Time and Motion

As you can see in the diagram, when you move through the field at a certain speed, it might seem as though you are traveling backward. Einstein also brought up that time is not linear but a perspective unique to each object. When moving across the Cartesian field, each object exchanges time and motion, and the time and motion of an object are the summation of the two. Consequently, each object experiences time differently.

According to Einstein's logic, if you travel from point "A" to point "B," which are 60 miles apart, and it takes you one hour, you would say you traveled at 60 mph. If you wanted to get there faster and doubled your speed, it would take you less time. A reasonable person might predict that time would reduce proportionally to the increase in speed. However, the idea that increasing speed would increase time is counterintuitive and does not hold in our observable universe.

Einstein's second premise states that if you continue increasing your speed, there will come a point when you approach zero time left. This concept is fundamental: as an object moves faster, time diminishes until it theoretically reaches zero. When an object has zero time left, its time has expired. This mass, often discussed in scientific circles, is the very fabric of the universe and has the potential to replenish itself. Einstein proposes that accelerating an object to incredible speeds has a similar effect to freezing it. Since time is perpendicular to motion and inversely proportional, both are finite. Pulling one end of this relationship at a 90-degree angle causes the other end to diminish.

Scientists sometimes overlook the fact that the universe is not void of anything. It is filled with instances of reality all around us, some observable and measurable, and others beyond our current capabilities to detect. Consider this: the sun is 93 million miles away, yet we can feel its heat. How does this heat propagate, and through what medium? It's clear that space cannot be a vast emptiness of nothingness, as propagation requires a medium.

Immanuel Kant was right when he suggested that our sensory limitations constrain our understanding. Our brains evolved to survive on Earth, focusing on essential sensors for survival rather than celestial sensors. There is undoubtedly more to the universe than we currently know or can detect. For instance, the Hubble Space Telescope, launched into low Earth orbit in 1990, captures light from distant galaxies. This light and other cosmic signals do not propagate through a vacuum but through the medium of space.

In conclusion, Einstein's theory shows that time and motion are deeply interconnected and unique to each object. As objects move, they use up their energy of time and motion. Understanding this relationship helps us appreciate the complexity of the universe and the genius of Einstein's insights. The universe is a dynamic interplay of time and motion, far from the empty void we might have once imagined. Through tools like the Hubble Space Telescope, we continue to explore and unravel the mysteries of our vast cosmos.

What happened at zero time?

Chapter 42.0: The Enigma of Zero Time

What happens at zero time? At zero-time, transportation is instantaneous. You either die or get transported instantaneously. At this moment, an object would return to a singularity state. This is akin to the concept in Star Trek where you can be everywhere simultaneously. However, even if you survive traveling at such speed, there are further perils to consider.

According to Einstein, light travels at a certain speed and behaves as both a wave and a particle. Some of these particles make up the human body. Now, let's consider the damaging effects of traveling faster than the speed of light on the human body. Imagine your body vibrating at the speed of light. This means that particles in your body are vibrating at that incredible speed. Traveling faster than light means vibrating faster than this internal vibration, which can be highly destructive.

Dimensional Formula of Speed

The dimensional formula of speed is given by:

$$[M^0 L^1 T^{-1}]$$

Where:

- M = Mass

- L = Length

- T = Time

Derivation

Speed (S) is defined as:

$$S = \frac{\text{Distance}}{\text{Time}}$$

The dimensional formula of distance is:

$$[M^0 L^1 T^0]$$

And the dimensions of time are:

$$[M^0 L^0 T^1]$$

On substituting these into the formula for speed, we get:

$$[S = \frac{[M^0 L^1 T 0]}{[M0 L^0 T^1]} = [M^0 L^1 T^{-1}]]$$

Therefore, speed is dimensionally represented as:

$$[M^0 L^1 T^{-1}]$$

When you move at an immense speed, such as the speed of light, your body's internal particles vibrate at that speed. If you move even faster, the vibrations increase beyond the body's natural capacity, causing potential disintegration or extreme stress on your molecular structure.

The Conundrum of Zero Time

At zero time, an object would experience instantaneous transportation. This is because at speeds approaching or exceeding the speed of light, time dilation becomes so extreme that time effectively stands still for the object in motion. In theoretical terms, reaching zero time means returning to a state of singularity, where the boundaries of time and space converge.

Einstein's theory of relativity suggests that time and motion are deeply interconnected and unique to each object. As objects move, they use up their "battery" of time and motion. This relationship emphasizes the dynamic and interconnected nature of the universe.

Universal Medium

It's important to understand that the universe is not a void of nothingness. It is filled with instances of reality in various forms, some of which we can observe and measure, while others remain beyond our detection. For example, the sun is 93 million miles away, yet we feel its heat. This heat propagates through a medium, not through empty space.

Our sensory limitations, as noted by Immanuel Kant, constrain our understanding. Our brains developed to survive on Earth, focusing on essential sensors rather than celestial ones. However, the universe comprises more than we currently understand.

The Hubble Space Telescope, launched into low Earth orbit in 1990, captures light from distant galaxies. These light and other cosmic signals propagate through the medium of space, not through a vacuum.

Conclusion

In summary, the concept of zero time introduces the idea of instantaneous transportation or returning to a singular state. The interplay between time and motion at such extreme speeds highlights the complexities of Einstein's theories. Understanding these relationships enriches our comprehension of the universe's dynamic and interconnected nature. The universe, far from being an empty void, is a vibrant and intricate tapestry of time, motion, and reality.

What shape and dimension is the universe?

Chapter 43.0: The Shape and Dimensions of the Universe

What shape is the universe? To explore this question, we must delve into dimensional physics, the study of the universe's dimensions. Classical physics describes three physical dimensions: from any given point in space, the fundamental directions are up/down, left/right, and forward/backward. Any movement can be expressed using these three dimensions.

In mechanics, the fundamental dimensions are time (t), mass (m), and length (l). Electromagnetism adds another fundamental dimension: electric charge (q).

Our Perception of Dimensions

We are 3D creatures living in a 3D world, but our eyes can only show us two dimensions. The depth we perceive is a trick our brains have learned, a byproduct of evolution positioning our eyes in front of our faces. To prove this, try closing one eye and playing tennis—you'll find it much harder to judge distances.

Beyond the Visible Dimensions

Classical physics tells us about the three spatial dimensions, but modern theories propose even more. For instance, one type of gluon chain behaves in four-dimensional space-time as the graviton, the fundamental quantum particle of gravity. In this view, gravity in four dimensions emerges from particle interactions in a gravity-less, three-dimensional world.

The Ten Dimensions of Reality

To explain the universe, we must consider multiple dimensions. Let's start at the beginning and explore the ten dimensions of reality:

1. Length

2. Width

3. Depth

4. Time

5. Probability (Possible Universes)

6. All Possible Universes branching from the same start conditions

7. All Possible Spectrums of Universes with different start conditions

8. The Fundamental Constants of Nature

9. The Laws of Physics

10. The Ultimate Existence beyond our perception

Ten-Dimensional Beings

What is a ten-dimensional being? Such an entity would be one with everything, with an essence present in every form of existence. In theoretical terms, a ten-dimensional being could be likened to Brahman in Hindu philosophy—a single, timeless, infinite entity encompassing everything.

The Dimensions According to String Theory

According to string theory, the universe operates with ten dimensions. We are familiar with three spatial dimensions (length, width, depth) and one temporal dimension (time). String theory suggests the existence of additional dimensions beyond our immediate perception, shaping the universe in ways we can't directly observe.

Higher Dimensions and Practical Applications

Time is often considered the fourth dimension, essential for understanding motion through space. Under Einstein's relativity, motion through space and time are intertwined.

In wellness, Swarbrick's '8 Dimensions' model provides a practical framework for holistic health:

1. Physical

2. Spiritual

3. Social

4. Intellectual

5. Emotional/Mental

6. Occupational

7. Environmental

8. Financial

The 11th Dimension and Beyond

The 11th dimension, proposed by Superstring Theory, addresses questions about the fabric of space-time, involving nine dimensions of space and one dimension of time. This expands our understanding of the universe's complexity.

The 26 Dimensions of Closed Unoriented Bosonic String Theory

Closed Unoriented Bosonic String Theory posits 26 dimensions, explained through the traceless Jordan algebra J3(O)oJ_3(O)_o of 3x3 Octonionic matrices. This model interprets the 26 dimensions as encompassing 4-dimensional physical space-time plus additional dimensions that describe the universe's intricate structure.

Conclusion

In summary, the universe's shape and dimensions extend far beyond our three-dimensional perception. By exploring theories like string theory and higher dimensions, we gain insight into the complex and fascinating structure of our reality. The universe is a dynamic interplay of dimensions, each contributing to the vast tapestry of existence. Understanding these dimensions enhances our appreciation of the cosmos and the underlying principles that govern it.

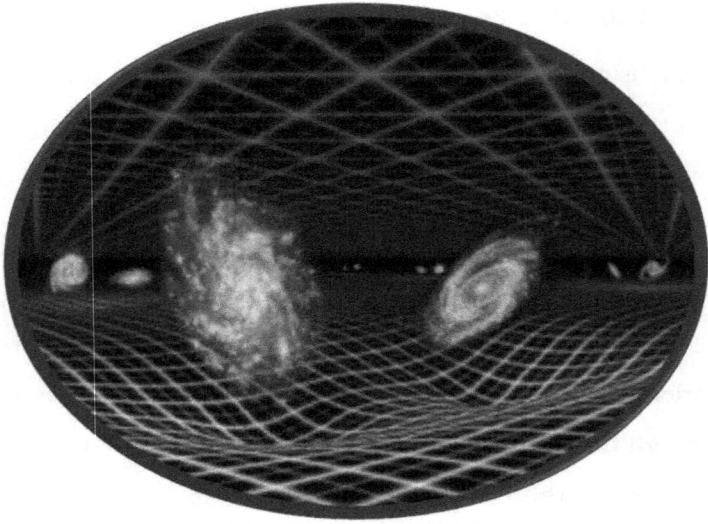

GERALD CLERGE

The possible likeness of our universe?

Chapter 44.0: The Shape of the Universe

I believe the universe closely resembles an egg—more oval than anything else. While anyone can imagine the universe's shape in various forms, dimensional physics encourages us to think like children, imagining the unseen dimensions of our cosmos.

The Basics of Dimensional Physics

In geometry, a point exists without length, width, shape, or size; it only has a position. We call this point an "instant of reality." Although we do not fully understand it, we know it exists. Let's consider this point as a singularity. A singularity is a point where a given mathematical object ceases to be well-defined, lacking differentiability or analyticity. From this singularity, various complexities arise, forming an oval, three-dimensional universe.

Our Perception of the Universe

As a species, we have determined our universe to be three-dimensional. This structure is a three-dimensional representation of what might otherwise be seen as a two-dimensional object. Some dimensional physicists might argue that the universe's dimension is akin to a cube.

Others might propose that the universe is shaped like a tetrahedron. A tetrahedron is the three-dimensional case of a more general Euclidean simplex concept and may be called a 3-simplex. With four triangular faces, a tetrahedron resembles a "triangular pyramid."

Alternatively, the universe could be likened to an orthotope, a hyperrectangle, or a three-dimensional object. An orthotope, also known as a rectangular prism, rectangular cuboid, or parallelepiped, has n dimensions where all edges have equal length, forming an n-cube.

The Concept of a Tesseract

Some dimensional scientists suggest moving beyond the three-dimensional view to consider a four-dimensional universe, represented by a tesseract. In geometry, a tesseract is the four-dimensional analog of a cube; just as a cube extends a square into the third dimension, a tesseract extends a cube into the fourth dimension. The surface of a tesseract consists of eight cubical cells, compared to a cube's six square faces.

Visualizing a Four-Dimensional Universe

Analyzing a four-dimensional universe, we see it shaped like a tesseract. A four-dimensional cube has eight sides, compared to a regular cube's six. Each of these eight sides consists of a three-dimensional cube, one dimension down from itself, similar to how each side of a three-dimensional cube is a two-dimensional square.

The Egg-Shaped Universe

Envisioning the universe as egg-shaped allows us to explore the idea of dimensions creatively. The oval shape might symbolize the curved, expanding nature of the cosmos. This visualization helps us grasp complex theoretical concepts through more relatable, tangible forms.

Conclusion

The universe's shape remains a subject of debate and imagination. While dimensional physics offers various models—from cubes and tetrahedrons to orthotopes and tesseracts—the true form of the universe may elude precise definition. By considering different dimensions and shapes, we gain a deeper understanding of the complexities and wonders of our cosmos. Whether we envision the universe as egg-shaped, cubical, or a tesseract, each perspective enriches our comprehension and sparks further curiosity about the nature of reality.

Chapter 44.1: The Shape of the Universe

If the universe has a shape like a tesseract, it could theoretically contain two-dimensional, three-dimensional, and four-dimensional beings. However, as three-dimensional beings, we would not fully perceive the fourth dimension with our current brain structure. The concept of a four-dimensional universe challenges our understanding and requires us to expand our imagination.

The Pentachoron and Higher-Dimensional Shapes

The universe might look like a pentachoron. In geometry, the 5-cell is the regular convex 4-polytope (four-dimensional analog of a Platonic solid) with the Schläfli symbol {3,3,3}. It is a four-dimensional object bounded by five tetrahedral cells, also known as a C5, pentachoron, pentatope, pentahedron, or tetrahedral pyramid. The pentachoron is a four-dimensional pyramid with a tetrahedral base, analogous to the tetrahedron in three dimensions and the triangle in two dimensions.

Other shapes include a tetrahedron with four triangular faces, a cube with six square faces, an octahedron with eight triangular faces, a dodecahedron with twelve pentagonal faces, and an icosahedron with twenty triangular faces. The study of these shapes often involves 'regular polytopes' instead of 'Platonic solids' in higher dimensions. All faces of a regular polytope must be lower-dimensional regular polytopes of the same size and shape, and all vertices and edges must look identical, achieving maximal symmetry.

Visualizing Four-Dimensional Objects

To visualize these shapes, we start with familiar forms. A Platonic solid in ordinary three-dimensional space appears as a sphere with its surface divided into polygons. Similarly, a four-dimensional polytope looks like a sphere in four-dimensional space with its surface divided into polyhedra. A sphere in four-dimensional space is called a '3-sphere.' People living on its surface would experience it as a three-dimensional universe with the curious feature that if you travel straight in any direction, you eventually return to your starting point. This is similar to walking in a straight line on an ordinary sphere on Earth.

Concept of a Tesseract

A tesseract is the four-dimensional analog of a cube. Just as a cube extends a square into the third dimension, a tesseract extends a cube into the fourth dimension. The surface of a tesseract consists of eight cubical cells, while a cube's surface consists of six square faces. Understanding a tesseract helps us grasp the complex nature of higher-dimensional spaces.

Higher-Dimensional Spaces

Beyond the tesseract and pentachoron, higher-dimensional spaces introduce even more complex structures. In four dimensions, there are precisely six regular polytopes. Visualizing these shapes can be challenging, but they provide insights into the universe's fundamental structure.

Conclusion

The shape of the universe remains a fascinating and complex topic. Whether we envision it as an egg, a tesseract, a pentachoron, or other higher-dimensional shapes, each perspective offers unique insights into the nature of reality. By exploring dimensional physics and higher-dimensional objects, we expand our understanding of the cosmos and appreciate the intricate structures that make up our universe. The journey into higher dimensions reveals a universe far more complex and wondrous than our three-dimensional perception suggests, inspiring us to continue exploring and questioning the nature of existence.

Chapter 44.2: The Shape of the Universe and Dimensionality

If the universe looks like the image below, it suggests that all dimensions exist within a sphere. Let's explore this concept using the rich tapestry of geometry and dimensional physics.

Symmetrical Polygons in Two Dimensions

In two dimensions, the most symmetrical shapes are the 'regular polygons.' All the edges of a regular polygon are of equal length, and all the angles are equal. Regular polygons include the equilateral triangle, square, and regular pentagon, extending to an infinite number of regular polygons, each with n sides for n > 3.

Regular Polyhedra in Three Dimensions

In three dimensions, the most symmetrical shapes are the 'regular polyhedra,' also known as 'Platonic solids.' These solids have faces that are regular polygons of the same size, and their vertices look identical. The five Platonic solids are:

1. **Tetrahedron**: 4 triangular faces

2. **Cube (Hexahedron)**: 6 square faces

3. **Octahedron**: 8 triangular faces

4. **Dodecahedron**: 12 pentagonal faces

5. **Icosahedron**: 20 triangular faces

Regular Polytopes in Four Dimensions

In four dimensions, there are six regular polytopes. Visualizing these shapes involves imagining a 3-sphere (a sphere in four-dimensional space). Just as a Platonic solid looks like a sphere with its surface divided into polygons in three-dimensional space, a 4-dimensional polytope appears as a sphere divided into polyhedra in four-dimensional space.

Key Four-Dimensional Polytopes

1. **4-Simplex (Hyper Tetrahedron)**: 5 tetrahedral faces

2. **Tesseract (Hypercube)**: 8 cubical faces

3. **16-Cell (Hyper Octahedron)**: 16 tetrahedral faces

4. **120-Cell (Hyper Dodecahedron):** 120 dodecahedral faces

5. **600-Cell (Hyper Icosahedron):** 600 tetrahedral faces

6. **24-Cell:** 24 octahedral faces, unique to four dimensions with no lower-dimensional analog

Higher-Dimensional Spaces and Visualization

Visualizing these higher-dimensional polytopes requires breaking down their structure. A 3-sphere appears like an ordinary three-dimensional space, but it 'wraps around' at a distance. By focusing on a local portion of this space divided into polyhedra, we can conceptualize these complex shapes.

Practical Implications and the Universe's Structure

Understanding the shape and dimensions of the universe can provide insights into its fundamental nature. For instance, the 24-cell, unique to four dimensions, offers a model of our universe's structure. Exploring these shapes can help explain how the universe operates on different levels of dimensionality.

Conclusion

The shape of the universe, whether imagined as an egg, tesseract, pentachoron, or another higher-dimensional shape, opens up a world of possibilities. By delving into dimensional physics and geometry, we gain a deeper understanding of the cosmos. Each perspective, from regular polygons to four-dimensional polytopes, enriches our comprehension of the universe's complexity and beauty. Through these explorations, we continue to push the boundaries of knowledge, uncovering the intricate patterns that define our reality.

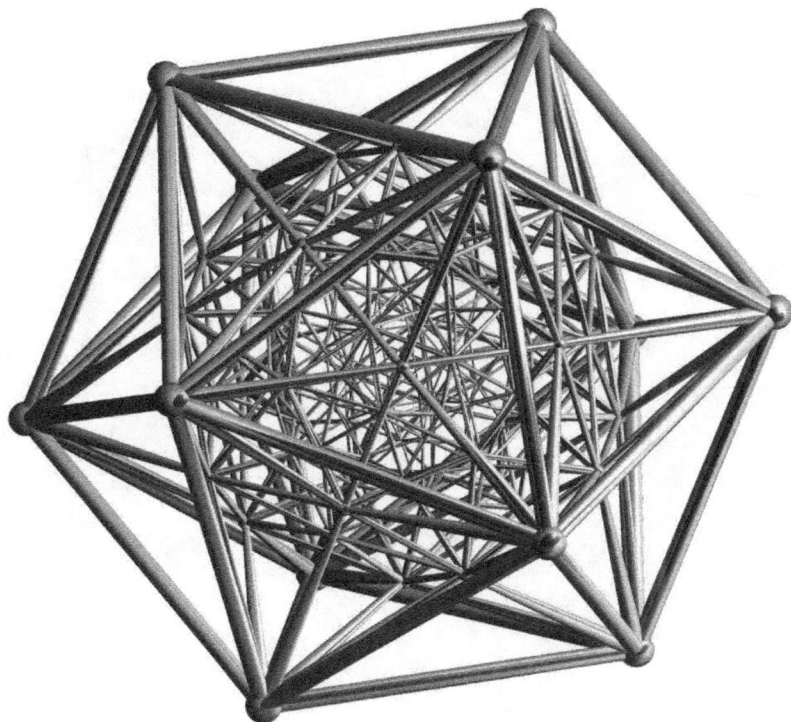

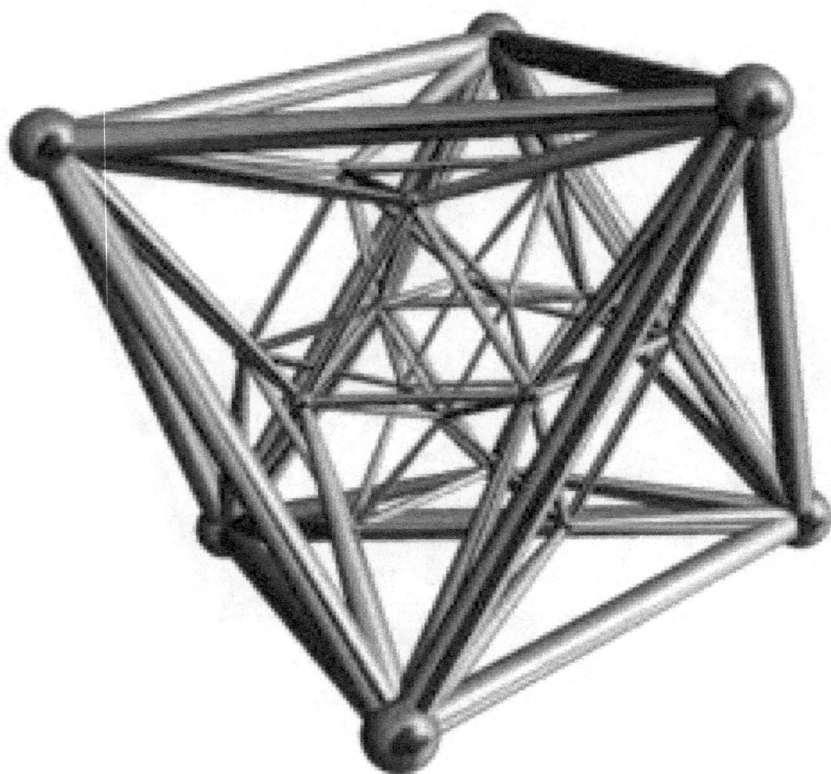

GERALD CLERGE

Chapter 44.3: The Complexity of the Universe and Dimensional Understanding

Now that we grasp the complexity of the universe and the concept of dimensions, let's reflect on our position as three-dimensional beings. As residents of one of many three-dimensional planes, our knowledge of higher dimensions is limited. How can we accurately depict the universe without understanding the other dimensions—first, second, and beyond?

Understanding Higher Dimensions

When a three-dimensional being looks at the universe, they perceive it from a 3D perspective. For us, the galaxy might resemble a Cartesian Coordinate system. However, a four-dimensional being would not see the cosmos in the same way.

The Concept of a Four-Dimensional Cube

A four-dimensional cube, or tesseract, has eight sides. Each of these sides is a three-dimensional cube, just as each side of a three-dimensional cube is a two-dimensional square. In four dimensions, a being would perceive length, width, depth, and an additional spatial dimension.

Perception of Higher Dimensions

A four-dimensional being could see you in your entirety, but you would be unable to see a four-dimensional being fully. Our brains are not programmed for four-dimensional perception, similar to how older operating systems may not work with new designs.

- **Speed:** A four-dimensional being would process information faster.
- **Memory:** A four-dimensional being would have more memory capacity.
- **Cores:** A four-dimensional being is likely to have more processing cores.

The software required to process a four-dimensional image does not exist in a 3D world. Our brains' processors would not have adequate memory to handle the resources and code needed to process such a being. We would need billions more processors to process a four-dimensional being. The graphics might suffer from stuttering, and there could be endless disk thrashing as virtual memory is swapped in and out.

The Implications of Higher Dimensions

Considering the 24 dimensions, we should hold off on proposing a theory of everything until we better understand these dimensions. Our current knowledge is limited to our three-dimensional experience.

Visualizing Higher Dimensions

Understanding higher dimensions involves imagining more complex structures:

- **Pentachoron (5-Cell):** A four-dimensional object bounded by five tetrahedral cells.
- **Tesseract (Hypercube):** An eight-sided four-dimensional cube.
- **16-Cell (Hyper Octahedron):** A polytope with 16 tetrahedral faces.
- **120-Cell (Hyper Dodecahedron):** Composed of 120 dodecahedral faces.
- **600-Cell (Hyper Icosahedron):** Composed of 600 tetrahedral faces.
- **24-Cell:** Unique to four dimensions with no lower-dimensional analog.

Conclusion

Given our understanding of the universe's complexity and dimensions, we recognize the limits of our three-dimensional perspective. Higher dimensions introduce concepts that challenge our perception and require advanced mental and technological frameworks. By exploring these dimensions, we expand our knowledge and appreciate the universe's intricate structure.

Our journey into higher dimensions reveals a universe far more complex than our three-dimensional view suggests. Whether visualizing a tesseract, pentachoron, or other higher-dimensional shapes, each perspective deepens our understanding of the cosmos. This ongoing exploration inspires us to continue pushing the boundaries of knowledge, uncovering the mysteries of existence.

What is an object's dimension, and what is in the space-time continuum?

Chapter 45.0: Exploring Dimensions and the Space-Time Continuum

In physics and mathematics, dimensions play a crucial role in understanding the structure and behavior of space and objects within it. A dimension can be informally defined as the minimum number of coordinates required to specify any point within a space. For example, a line has one dimension because only one coordinate is needed to describe a point on it.

Dimensional Analysis

Dimensional analysis is a method used in engineering and science to analyze the relationships between physical quantities by identifying their base quantities and units of measure. It involves tracking these dimensions as calculations or comparisons are performed. This technique is essential in converting between units, ensuring consistency in equations, and deriving relationships between different physical quantities.

Dimensional Analysis in Chemistry

In chemistry, dimensional analysis is the process of converting between units using conversion factors. It involves expressing related physical quantities in the desired units to facilitate comparisons and calculations. This method ensures that equations and measurements are accurate and consistent across different unit systems.

Dimensional Analysis in Physics

In physics, dimensional analysis is a powerful tool used to express physical quantities in their fundamental dimensions. This method is particularly useful when there is insufficient information to set up precise equations. By breaking down complex equations into their basic dimensions, scientists can gain insights into the relationships between different physical variables.

Dimensional Analysis in Mathematics

Dimensional analysis, also known as the Factor-Label Method or the Unit Factor Method, is a problem-solving technique in mathematics that leverages the fact that any number or expression can be multiplied by one without changing its value. This method is valuable in simplifying complex problems and ensuring that calculations are consistent and accurate.

Dimensional Analysis in Fluid Mechanics

In fluid mechanics, dimensional analysis is used to predict physical parameters that influence fluid behavior, heat transfer, and thermodynamics flow. This analysis involves the fundamental units of mass, length, and time (MLT), allowing engineers and scientists to understand and predict the behavior of fluids under various conditions.

Rayleigh's Method of Dimensional Analysis

Rayleigh's method, named after Lord Rayleigh, is a conceptual tool used in physics, chemistry, and engineering to express functional relationships between variables in the form of an exponential equation. The method involves the following steps:

1. **Gather Independent Variables:** Identify all independent variables likely to influence the dependent variable.

2. **Functional Equation:** If R is a variable that depends on independent variables R1, R2, R3, ..., Rn, the functional equation can be written as R = F(R1, R2, R3, ..., Rn).

3. **Express in Exponential Form:** Write the equation in the form $R = C * R1^{a*}\ R2^{b*}\ R3^{c}\ ...\ Rn^{m}$, where C is a dimensionless constant, and a, b, c, ..., m are arbitrary exponents.

4. **Express in Base Units:** Express each quantity in the equation in the base units required for the solution.

5. **Dimensional Homogeneity:** Use dimensional homogeneity to obtain a set of simultaneous equations involving the exponents a, b, c, ..., m.

6. **Solve Equations:** Solve these equations to determine the values of the exponents a, b, c, ..., m.

7. **Form Non-Dimensional Parameters:** Substitute the values of the exponents into the central equation and form non-dimensional parameters by grouping variables with like exponents.

Drawbacks

One drawback of Rayleigh's method is that it doesn't provide information regarding the number of dimensionless groups obtained through dimensional analysis. Despite this limitation, the method remains a valuable tool for understanding the relationships between physical variables.

By expanding on these topics, we can gain a deeper understanding of dimensions and their critical role in various scientific and engineering fields. Dimensional analysis provides a robust framework for solving complex problems and ensuring the accuracy and consistency of calculations.

Chapter 45.1: Understanding Dimensions - From Dots to the Space-Time Continuum

The Basic Structure: The Point

To understand dimensions, we begin with the most fundamental concept: the point. A point is a 0-dimensional mathematical object, a specific location in space that requires an n-tuple of coordinates for its precise specification in n-dimensional space. In higher dimensions, points often become synonymous with vectors, leading to the term n-vectors for points in n-dimensional space.

The Next Step: The Line

In geometry, a line is defined as a straight, one-dimensional figure with no thickness, extending infinitely in both directions. It represents the shortest distance between any two points. Lines serve as the building blocks for more complex shapes and figures, guiding the eye and imparting structure to visual representations.

Shapes and Figures

A shape or figure refers to the form of an object or its external boundary, outline, or surface. Unlike properties such as color or texture, shapes are defined purely by their geometric properties. Lines and shapes act as visual cues, directing our attention and activating our knowledge. They do not need to be straight or contiguous; even a broken or curved line can guide the eye and convey meaning.

From Zero Dimensions to Infinite Possibilities

Imagine a scenario where something emerges from nothingness—a point in space. Picture our planet suspended in outer space, with reality hanging in the void. In this void, there is no time, no dimensions, no movement, no directions. Zero dimensions represent a state with no width, length, or height. Yet, it holds an intriguing duality: simultaneously the smallest and the largest it can be.

When this instant of reality extends into a straight line, we get what is known as a ray. A ray consists of different points, has one endpoint, and stretches infinitely at the other end. This seemingly simple concept of a ray allows us to create distant objects, all of zero dimension, since a line itself has one side of zero dimension and a zero-dimension point at each end.

Building Blocks of Dimensions

1. **The Point (0D):** A singular location with no dimensions.
2. **The Line (1D):** An infinite collection of points extending in one direction, with no width or height.
3. **Shapes in 2D:** By connecting multiple lines, we create shapes with two dimensions, having both length and width but no depth.
4. **Objects in 3D:** Adding another dimension to 2D shapes results in three-dimensional objects, possessing length, width, and height.

Visualizing Higher Dimensions

While we can easily visualize 1D, 2D, and 3D objects, higher dimensions challenge our comprehension. A hypothetical cube in the first dimension, for example, would appear as a line equal in length to the cube but devoid of width or height.

By exploring these fundamental concepts, we gain insights into the fascinating world of dimensions and the space-time continuum. Each step, from points to lines to complex shapes, forms the foundation for understanding the intricate structure of our universe.

Chapter 45.3: From Dots to Dimensions

Moving from Zero Dimension to Two Dimensions

Imagine looking at a dot again, and this time you see another line stretching away from it. Transforming a line segment in a direction perpendicular to the 1-dimensional direction brings you into the second dimension. Now the object has a coordinate and two sizes—

length and width. These two reference points allow you to create two-dimensional objects.

Two-Dimensional Shape: Initial Definition

A two-dimensional shape is one that has length and width but no depth. In mathematics, these shapes often correspond to objects with common geometric attributes in the real world. For instance, a cube in three dimensions would exist as a square in the second dimension. You can draw a representation of a 3D cube on a 2D plane, but this depiction does not convey the true essence of a 3D cube. Instead, it represents the third dimension superimposed on the second.

The Transition to Three Dimensions

Now, imagine reality extending further with another ray. This third ray adds depth to our object, transforming it into a three-dimensional object. The three rays extending away from the point give this object depth, making it a three-dimensional object. In three dimensions, we have three vantage points to define an object: length, width, and depth.

Three-Dimensional Objects: Examples

The objects around you—the ones you can pick up, touch, and move around—are three-dimensional. These shapes have a third dimension: depth. Examples of three-dimensional objects include cubes, prisms, pyramids, spheres, cones, and cylinders. Each of these objects is characterized by its three-dimensional structure, allowing for physical interaction and spatial presence.

Visualizing the Transition

To understand this transition, imagine the following steps:

1. **Point (0D):** A singular location in space with no dimensions.
2. **Line (1D):** Extending the point in one direction creates a line with length but no width or height.
3. **Plane (2D):** Extending the line perpendicularly in another direction creates a plane with length and width but no depth.
4. **Space (3D):** Adding depth by extending the plane perpendicularly in a third direction creates a three-dimensional space.

By visualizing these steps, we can appreciate the progression from zero dimensions to three dimensions. Each additional dimension adds a layer of complexity and richness to our understanding of the physical world.

Let's delve deeper into the properties of these dimensions and explore their significance in various fields of study. Whether it's in mathematics, physics, or everyday life, dimensions play a fundamental role in shaping our perception of reality. Shall we continue exploring the fascinating world of dimensions and the space-time continuum?

Chapter 45.4: The Leap to the Fourth Dimension

Extending Beyond Three Dimensions

Imagine once again looking at a dot, but this time you see another ray shooting out of the point. With this addition, the universe gains a fourth vantage point to create objects. This new dimension adds complexity and depth to our understanding of space.

Understanding Four-Dimensional Space

Four-dimensional (4D) space is a mathematical extension of the three-dimensional (3D) space we are familiar with. In everyday life, we only need three numbers, or dimensions, to describe the sizes or locations of objects. However, in 4D space, we add an extra dimension, creating a richer and more complex spatial structure.

The Tesseract: A Glimpse into 4D

A tesseract, also known as a hypercube, is a four-dimensional mathematical object. Like a cube in 3D space, a tesseract has lines of equal length at right angles. It is the extension of a square into four-dimensional space, similar to how a cube extends the notion of a 2D square into three dimensions.

Moving Through Dimensions

1. **Second Dimension (2D):** In 2D, we have length and width, allowing for basic shapes and geometry.
2. **Third Dimension (3D):** Moving into the third dimension, a 2D square gets extruded in a perpendicular direction to both of its sides, creating a cube. In Cartesian terms, this involves the X and Y directions for the square, and the Z direction for the cube. Our traditional understanding of a cube includes measurements of width, length, and height.
3. **Fourth Dimension (4D):** In 4D space, a tesseract is formed by extending a cube into a fourth direction, perpendicular to the three spatial directions we know. This extension creates a hypercube, a complex structure with an additional dimension.

Visualizing the Fourth Dimension

Visualizing the fourth dimension can be challenging because our brains are wired to perceive only three dimensions. However, we can use analogies and mathematical models to understand the concept:

1. **Square to Cube:** In 2D, a square exists with length and width. In 3D, this square is extruded into a new direction to form a cube.

2. **Cube to Tesseract:** Similarly, in 4D space, a cube is extruded into an additional dimension, creating a tesseract.

Complexity of Higher Dimensions

As we move into higher dimensions, the math becomes more complex. Each additional dimension adds a layer of intricacy to our understanding of space. While basic shapes and geometry are relatively straightforward in 2D, the third dimension introduces depth and more complex spatial relationships. In 4D, these relationships become even more intricate, challenging our perceptions and mathematical tools.

By exploring these concepts, we can expand our understanding of the universe and the mathematical structures that describe it. The journey from points to lines to shapes, and eventually to higher dimensions, offers a fascinating glimpse into the complexity and beauty of the spatial world.

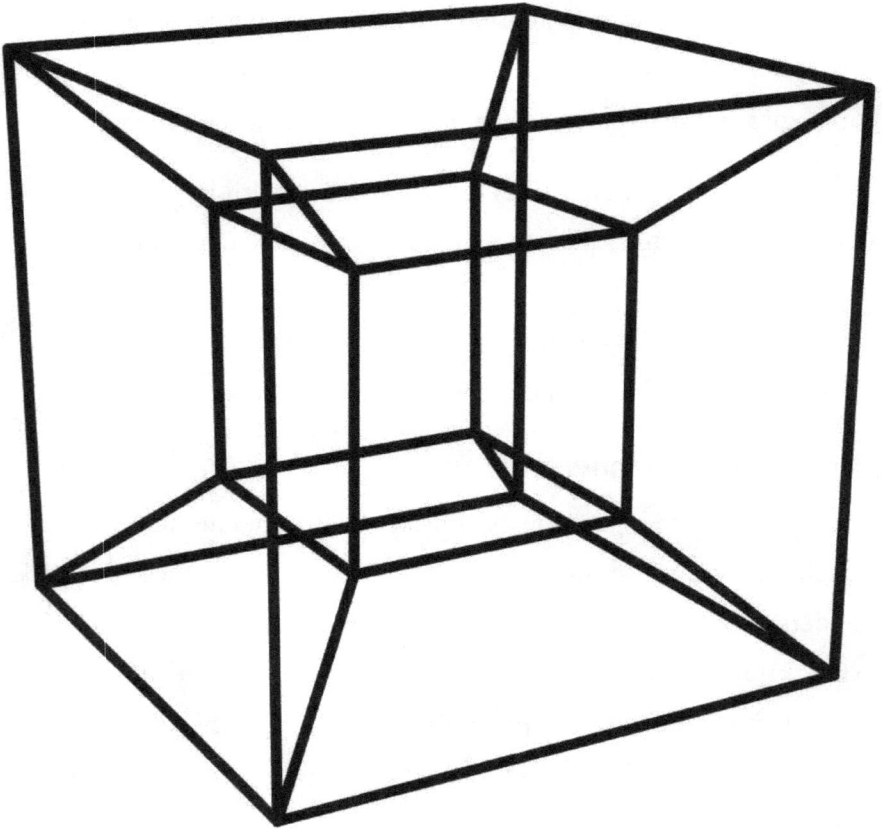

GERALD CLERGE

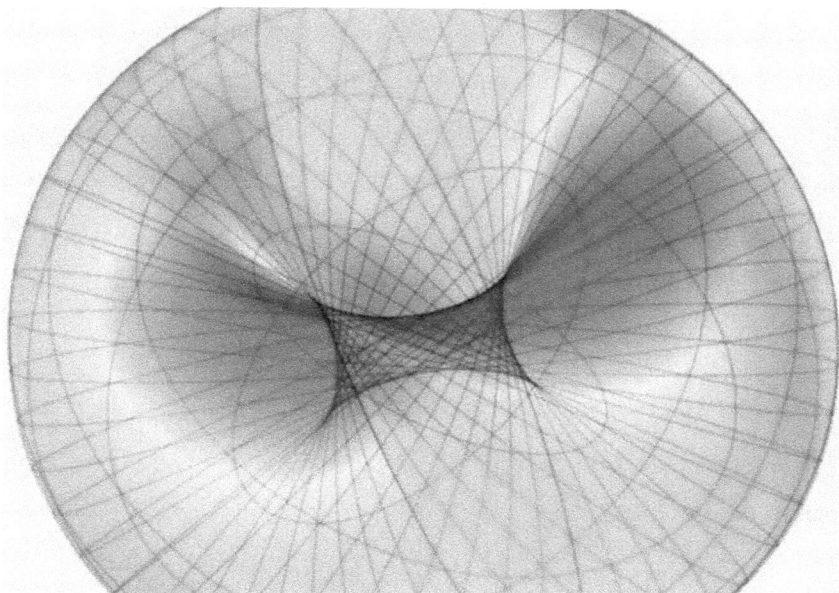

Chapter 45.5: Venturing into the Fourth Dimension

Understanding the Fourth Dimension

A fourth dimension is a realm where you can travel in a direction perpendicular to the three familiar spatial dimensions. To the untrained eye, this concept might seem perplexing. How could there be a direction that is perpendicular to a three-dimensional space? To grasp this idea, we need to explore dimensions systematically and observe the gradual changes that occur as we move through each one.

From Cubes to Tesseracts

In four-dimensional space, cubes transform into objects known as tesseracts. While objects in three dimensions have length, width, and height, objects in 4D possess an additional dimension called "trength." This extra dimension adds a layer of complexity and richness to the object's structure.

When we superimpose trength on any of the previous dimensions, it gives the object in the subsequent dimension a trength of 0 or an infinitesimally small value. All edges of a tesseract are of equal length, and all angles are right angles. Although this makes theoretical sense, visualizing a tesseract is challenging because our minds are accustomed to three dimensions. We need to project the fourth-dimensional object into the third dimension to conceptualize a tesseract.

Visualizing Four-Dimensional Space

To better understand the fourth dimension, let's break down the process step by step:

1. **Point (0D):** A singular location with no dimensions.
2. **Line (1D):** Extending a point in one direction creates a line with length but no width or height.
3. **Plane (2D):** Extending a line perpendicularly creates a plane with length and width but no depth.
4. **Space (3D):** Adding depth to a plane by extending it perpendicularly creates three-dimensional space, with length, width, and height.
5. **Hypercube (4D):** Extending a cube perpendicularly in a new direction creates a tesseract, a four-dimensional object with length, width, height, and trength.

The Complexity of Higher Dimensions

As we move into higher dimensions, the mathematics and visualizations become increasingly complex. While we can easily understand and draw objects in 2D and 3D, representing 4D objects requires us to stretch our imagination and mathematical tools. A tesseract, for instance, is a four-dimensional analog of the cube, with all its edges equal and all its angles right angles.

Practical Implications of the Fourth Dimension

Exploring the fourth dimension isn't just an abstract exercise; it has practical implications in various fields, including physics, computer graphics, and theoretical mathematics. By understanding higher dimensions, scientists and researchers can develop new theories and technologies that push the boundaries of our current knowledge.

For example, in theoretical physics, concepts like string theory and quantum mechanics often involve higher dimensions to explain complex phenomena that cannot be described using only three spatial dimensions. In computer graphics, the mathematics of higher dimensions can help create realistic 3D renderings and simulations.

Bridging the Gap Between Dimensions

To visualize a tesseract in 3D, we often use projections or cross-sections, similar to how a shadow or a slice of a 3D object can represent its 2D aspects. These visual tools help us bridge the gap between our familiar three-dimensional world and the abstract realm of four-dimensional space.

By continuing to explore the concepts of higher dimensions, we can expand our understanding of the universe and unlock new possibilities in science and technology. The journey from points to lines to shapes and beyond offers a fascinating glimpse into the complexity and beauty of the spatial world.

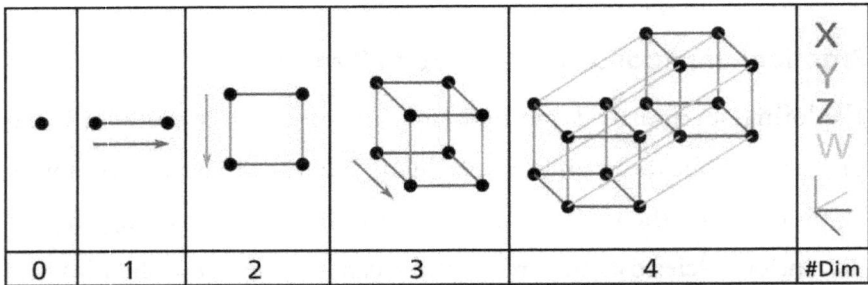

The universe must be multi-dimensional for it to exist.

Chapter 46.0: Understanding Dimensions Through the Human Body

Multi-3D Shapes in the Human Body

To fully grasp the concept of dimensions, we can use the human body as an example of a collection of multi-3D shapes. The human body is made up of various parts, each with its unique three-dimensional form. From the arrangement of molecules and atoms to larger elements, every component of your body is structured in different 3D shapes. These individual shapes come together to form the complex and integrated structure of the human body.

The Universe of Shapes and Dimensions

Similarly, when we break down the universe into its smallest and largest components, we find a diverse array of shapes with various dimensions. For instance, imagine slicing a part of the universe into a cube. Within this cube, you might see a structure resembling a board resting on two bricks. This scenario represents two three-dimensional areas with a one-dimensional space on top, separated by nothingness.

The area resembling a board is like a slit in the nothingness, with no dimensions on top or bottom. An actual thin slit has zero dimensions, with two endpoints also having zero dimensions. When a two-dimensional area appears on top of a one-dimensional space, it is reminiscent of the Phantom Zone from Superman's stories. In this fictional universe, the Phantom Zone is a two-dimensional prison where General Dru-Zod was exiled.

Exploring Higher Dimensions

When considering higher dimensions, think of the human mind as an example. Just as the mind adds a layer of complexity to our understanding, the universe, when broken down, reveals shapes with various dimensions. These dimensions come together to create unique spatial structures.

For example, envision an area of the universe with multiple three-dimensional locations. By adding two-dimensional and four-dimensional envelopes around these structures, we form a distinctive dimensional space. This multi-dimensional space is akin to combining different dimensional aspects to create a more complex and nuanced universe.

The Phantom Zone Analogy

In the Superman stories, the Phantom Zone is a two-dimensional prison where criminals are exiled. This concept can help us understand how higher dimensions can interact with lower dimensions. Just as the Phantom Zone exists as a separate dimensional space, we can imagine areas of the universe with varying dimensions interacting to create unique structures.

By exploring these concepts, we gain a deeper understanding of the complexity and beauty of dimensions. From the human body to the vast universe, dimensions play a fundamental role in shaping our reality.

Chapter 46.1: The Universe of Dimensions

Embracing the Boundless Dimensions of the Universe

Think of it this way: there is no design or dimension that the universe has not created. No human being can think outside the universe, so if you can dream it, it already exists in some form. Just as a human being is the sum of its three-dimensional shapes, the universe is the sum of its multi-dimensional shapes.

The Human Body: A Model of Multi-Dimensional Shapes

Consider the human body, which comprises various organs of different shapes and sizes. Each organ can be thought of as a distinct area in space, containing materials such as compound molecule chains, each with a unique structure. These structures come together to form the organ. For example, the kidney is an area in the body made up of various molecular structures. The concept of higher consciousness or dimension can be likened to the kidney, and the kidney itself is part of an even higher dimension—the human body.

A Universe of Diverse Dimensions

When we think of the universe, we can break it down into its smallest and largest components, revealing a plethora of shapes with various dimensions. Imagine slicing a section of the universe into a cube. Within this cube, you might see a structure resembling a board resting on two bricks, representing two three-dimensional areas with a one-dimensional space on top, separated by nothingness.

This area, resembling a board, is like a slit in nothingness, with zero dimensions on top and bottom. An actual thin slit has zero dimensions, with two endpoints also having zero dimensions. When a two-dimensional area appears on top of a one-dimensional space, it mirrors the concept of the Phantom Zone from Superman's tales. In this fictional universe, the Phantom Zone is a two-dimensional prison.

Higher Dimensions and Unique Spaces

Considering higher dimensions, the mind offers a fascinating analogy. Just as the mind adds complexity to our understanding, the universe, when broken down, reveals shapes with various dimensions that come together to form unique spatial structures. For instance, envision an area of the universe with multiple three-dimensional locations. By adding two-dimensional and four-dimensional envelopes around these structures, we create a distinctive dimensional space.

The Phantom Zone Analogy

In Superman's stories, the Phantom Zone is a two-dimensional prison where criminals are exiled. This concept helps us understand how higher dimensions can interact with lower dimensions. Just as the Phantom Zone exists as a separate dimensional space, we can imagine areas of the universe with varying dimensions interacting to create unique structures.

By exploring these ideas, we gain a deeper understanding of the complexity and beauty of dimensions. From the human body to the vast universe, dimensions play a fundamental role in shaping our reality.

Chapter 47.0: The Block Universe Theory and the Nature of Time

The Block Universe Theory

According to the block universe theory, the universe can be thought of as a giant block containing everything that has ever happened, is happening, and will happen. In this view, the past, present, and future exist equally, creating a static "block" of space-time. This concept challenges our conventional understanding of time as a linear progression from the past to the future.

The Paradox of Reality

However, this idea seems paradoxical and defies our intuitive understanding of reality. If everything existed simultaneously, it would resemble a still picture with no movement or motion. For motion to occur, there must be a distinction between different moments in time— an end and a beginning. The notion that the past, present, and future are all happening simultaneously implies that only one moment in time exists, has ever existed, or will ever exist. This concept appears self-defeating and contradicts our everyday experience of time.

Einstein's Perspective on Time

Albert Einstein once wrote, "People like us who believe in physics know that the distinction between past, present, and future is only a persistent illusion." In other words, Einstein suggested that time is an illusion. Many physicists have shared this view, proposing that actual reality is timeless and that our perception of time is merely a construct of our consciousness.

Timeless Reality

The idea that reality is timeless aligns with certain interpretations of modern physics. In these interpretations, time is not a fundamental aspect of the universe but rather an emergent property that arises from the interactions of physical entities. This perspective challenges our traditional understanding of time and suggests that the universe may be far more complex than we can perceive.

Reconciling the Block Universe with Our Experience

While the block universe theory offers a fascinating perspective, it also raises questions about how we experience time. Our perception of time as a continuous flow from past to present to future is deeply ingrained in our consciousness. Reconciling this perception with the idea of a timeless reality requires us to rethink the nature of time and our place within the universe.

The Implications of a Timeless Universe

Exploring the implications of a timeless universe can lead to profound insights about the nature of existence. If time is indeed an illusion, then our understanding of cause and effect, change, and progress must also be reconsidered. This perspective can influence various fields, from physics to philosophy, and reshape our understanding of reality.

By delving into these concepts, we can expand our knowledge of the universe and the fundamental nature of time. The block universe theory offers a thought-provoking framework for exploring the mysteries of existence and challenges us to rethink our conventional views of reality.

Chapter 47.1: The Structural Illusion of Reality

Is Reality Nothing But a Structural Illusion?

To explore this profound question, let's examine a single electron—the basic, yet mysterious, building block of the universe. The shape and composition of electrons are fundamental to understanding the nature of reality. Electrons are subatomic particles with a specific configuration representing their arrangement within orbital shells and subshells. The valence electrons, found in the outermost shell, play a crucial role in determining the unique chemistry of each element.

The Bohr Model of the Atom

The Bohr model provides a simplified visual representation of the atom. It depicts a central nucleus composed of protons and neutrons, with electrons orbiting in circular paths or shells around the nucleus. These electron shells or energy levels help us visualize the distribution of electrons within an atom.

The Composition of Atoms

We now understand that atoms consist of three primary subatomic particles: protons, neutrons, and electrons. These particles are themselves made up of even smaller constituents called quarks. Atoms were formed after the Big Bang, approximately 13.7 billion years ago, becoming the fundamental building blocks of matter in the universe.

Atomic Diagrams and Interactions

Atomic diagrams were developed to explain the interactions of elements on Earth and in space long before the direct observation of atoms was possible. These diagrams illustrate that the fundamental building block of matter is the atom, which consists of protons (with a positive electrical charge), neutrons (with no charge), and electrons (with a negative charge).

The Periodic Table

The periodic table is a tabular arrangement of chemical elements organized by their atomic number, from hydrogen (with the lowest atomic number) to oganesson (with the highest atomic number). The atomic number corresponds to the number of protons in the nucleus of an element's atom.

The Speed of Electrons

Electrons possess rest mass, meaning they cannot travel at the speed of light. While they can move incredibly fast, they are limited by the universe's ultimate speed limit—the speed of light in a vacuum, as established by Albert Einstein. Nothing with mass can travel faster than 300,000 kilometers per second (186,000 miles per second). Only massless particles, such as photons (which make up light), can achieve this speed.

Particles and Their Properties

In the physical sciences, a particle is a small, localized object attributed with several physical or chemical properties, such as volume, density, or mass. Particles come together to form the particulate matter that makes up everything in the universe.

By examining the structure and behavior of electrons and other subatomic particles, we gain a deeper understanding of the fundamental nature of reality. This exploration reveals that what we perceive as reality is indeed a complex interplay of various dimensions and structures, challenging our conventional notions and inviting us to reconsider the true essence of existence.

Chapter 47.2: The Theory of Subatomic Particles

Subatomic Particle Theory

In the physical sciences, a subatomic particle is a particle smaller than an atom. The interactions of these particles are explained within the framework of quantum field theory, which views these interactions as the creation and annihilation of quanta of corresponding fundamental interactions. This theory successfully blends particle physics with field theory.

Components of an Atom

Subatomic particles, smaller than atoms, form the foundational components of all matter. The three primary subatomic particles that make up an atom are protons, neutrons, and electrons. The nucleus, located at the center of the atom, contains protons and neutrons.

The Dual Nature of Electrons

Electrons, along with all other quantum objects, exhibit dual properties: they are partly waves and partly particles. More accurately, electrons are not traditional waves or particles; they are quantized fluctuating probability wavefunctions. This means their exact position and momentum are probabilistic rather than deterministic.

The electron cloud is the region surrounding an atomic nucleus where electrons are likely to be found. This negative charge region is associated with atomic orbitals and is defined mathematically as an area with a high probability of containing electrons.

The Role of Quarks

Quarks are fundamental particles that move at high speeds and constitute protons and neutrons, which in turn make up the nucleus of an atom. Each proton and neutron contains three quarks. There are different varieties of quarks, with protons and neutrons being composed of up quarks and down quarks.

The Motion and Energy of Subatomic Particles

Electrons, as basic building blocks of the universe, cannot move faster than light. If we consider being outside space and time, the concept of speed becomes irrelevant. Within the universe, electrons travel close to the speed of light, raising questions about what creates the fabric of the universe and where electrons get the energy to move through space and time. They are in perpetual motion, contributing to the dynamic nature of reality.

The Nature of Reality

To understand the essence of reality, let's consider the Bible's claim that in the beginning, there was nothing. Both science and the Bible assert that initially, there was nothing. According to scientific explanations, out of this nothingness, an existence emerged. This existence began to heat up and, in a single instant, exploded, giving birth to the universe. This event, known as the Big Bang, spread matter and energy into nothingness, creating everything we now know as the universe.

The universe can be viewed as a protected shield against nothingness, where everything that exists is contained. This perspective suggests that nothing within the created universe can exist outside of it.

By examining the properties and behaviors of subatomic particles, we gain insights into the fundamental nature of the universe. These particles, in perpetual motion, shape the fabric of reality and challenge us to rethink our understanding of existence.

Chapter 47.3: The Expanding Universe and the Nature of Time

The Discovery of an Expanding Universe

In the 1920s, astronomer Edwin Hubble made a groundbreaking discovery that the universe is not static but expanding. This revelation provided compelling evidence for the Big Bang theory, which posits that the universe was born from a colossal explosion. For a long time, scientists believed that the gravity of matter within the universe would eventually slow down this expansion.

The Nature of Nothingness and Expansion

The concept of nothingness remains elusive. If something travels into this void, one might assume that nothingness behaves like normal space and time. Alternatively, the force of the Big Bang might still be at work, pushing the universe outward in all directions and creating motion and time. In this context, time is linear, progressing from the moment of the Big Bang to the universe's eventual end. Time travel within this universe might be theoretically possible but only in one direction: the future.

The Linear Flow of Time

Think of time as a linear movement starting from the Big Bang's epicenter. To reverse time, one would have to go against the force of time, which theoretically has no opposition in the expanding universe. However, the notion of time flowing backward faces significant theoretical challenges.

The Potential End of Expansion

There may come a time when the force driving the expansion of the universe ceases, or the universe expands into oblivion, akin to all explosions. When this happens, the universe could cease to exist until another explosive event restarts the cycle. While this scenario seems plausible, it is based on current scientific observations, notably from the Hubble Space Telescope.

The Hubble Space Telescope and Observations

Launched into low Earth orbit in 1990, the Hubble Space Telescope has become one of the largest and most versatile space telescopes, providing vital research data and captivating public interest. However, the observable universe through Hubble represents only a fraction of the entire cosmos.

Examining the Limits of Observation

Scientists must make assumptions about the measurable size of the universe based on their observations. The portion of the universe observable through the Hubble Telescope may not represent the entirety or even a significant percentage of the universe. Despite Hubble's ability to see stars billions of light-years away, it remains possible that we observe only a small part of a larger universe that may be both contracting and expanding.

The Nature of the Universe

Science considers the universe to be malleable, composed of particles and waves. Light itself is made up of particles (photons) and waves. Protons and neutrons are constructed from quarks, while electrons remain fundamental particles not built from smaller components. In the physical sciences, a particle is a localized object with properties such as volume, density, or mass. The term particle is versatile and used across various scientific disciplines.

By exploring the concepts of an expanding universe, the nature of time, and the fundamental particles that compose everything, we gain insights into the complexity and vastness of our cosmos. The Hubble Space Telescope has significantly advanced our understanding, but much remains to be discovered.

Chapter 47.4: The Vastness of the Universe and Its Expansion

Contemplating the Immensity of the Universe

Imagine the vastness of the universe and our tiny position within it. It would be unreasonable to think that every signal passing through the universe passes through Earth or even comes close to it. It's akin to saying that every wave that creates the beauty of the United States passes through a single grain of sand in Nebraska. These giant waves of various frequencies form the landscape of the United States, and one grain of sand receives only a minute portion of these waves. Similarly, defining the entire universe based on a microscopic sampling is highly misleading.

Examining the Theory of Gravity Slowing Expansion

Let's consider the second flaw in the theory—that gravity slows the universe's expansion. Rewinding to the Big Bang, an explosion of matter and energy proliferated into the void, resulting in the universe's continuous expansion. According to this theory, the universe's mass would eventually slow its expansion. However, how can something have weight in nothingness? This poses a significant conceptual challenge.

The Cyclical Nature of the Universe

Science suggests that the universe expands, collapses, and repeats this cycle infinitely. However, it remains unclear whether the universe is reconstructed the same way each time. The expanding universe theory and my pulsating universe theory share similarities, but they differ in how they perceive expansion. In my theory, expansion and collapse occur much faster. Using scientific theories alone does not fully explain the formation of objects in the universe.

The Hubble Space Telescope and Observational Limits

The Hubble Space Telescope, launched into low Earth orbit in 1990, has provided invaluable insights into the universe. However, the portion of the universe observable through Hubble represents only a fraction of the entire cosmos. Assuming that the observed expansion is indicative of the entire universe's behavior might be misleading.

The Mallet Universe Theory

Science considers the universe to be malleable, composed of particles and waves. Light itself is made up of particles (photons) and waves. Protons and neutrons are constructed from quarks, while electrons remain fundamental particles not built from smaller components. In the physical sciences, a particle is a localized object with properties such as volume, density, or mass.

Challenges to the Expanding Universe Theory

One challenge to the expanding universe theory is that the observable portion might not represent the whole universe. Even though the Hubble Telescope can see stars billions of light-years away, it might be capturing only a small part of a larger universe that contracts and expands. This observation challenges the assumption that the entire universe behaves similarly to the observed portion.

By exploring these concepts, we gain a deeper understanding of the universe's complexity and vastness. The questions raised about the nature of expansion and the limitations of our observations remind us that much remains to be discovered.

Chapter 47.5: Reconciling Expansion, Time, and the Structure of Reality

Observing the Expansion of the Universe

A reasonable deduction from the expanding universe theory, one that might not be commonly noticed, highlights the fascinating behavior of celestial objects. Observations indicate that the universe's expansion speed is approximately 150,680.03 mph. The velocity at which distant galaxies recede from an observer continuously increases over time. Comparatively, light travels at 669,600,000 mph, and electrons move near light speed. Given this, if electrons travel at near light speed and galaxies move slower, electrons will eventually surpass the very universe they inhabit. This suggests that either the electron's momentum is not linear, or our universe is a constructed field of illusion, creating our perceived reality.

The Dual Nature of Time

Reconciling Einstein's theory of localized time with the expanding theory of linear centralized time presents challenges. If time is linear, how can it simultaneously be localized? According to Einstein, moving electrons create time, and their orbits, dictated by probability clouds, support the concept of localized time. This makes more sense than envisioning the universe as producing a linear time from the Big Bang to its end.

The Instantaneous Nature of the Big Bang

Consider the nature of the Big Bang, which happened instantaneously. Science often scoffs at religious claims that the universe was created in a day, yet the formation of the universe took less than a day according to scientific understanding. While the formation of various objects took time, the universe's inception was immediate. Before the universe, in the void of nothingness, time did not exist, and everything occurred at once.

Examining the Expanding Universe Theory

The explosion of matter into the void, marked by intense heat and subsequent cooling, led to the creation of the universe. This aligns with biblical narratives like Moses' Journal, which claims that God created the universe instantly. Believers often overlook the notion that time existed post-creation, and thus, subsequent events required time.

The Dynamic Stage of the Universe

If the universe is still expanding, we remain in the explosion stage, potentially closer to its end. Galaxies, composed of waves and particles, reveal the universe's structure. Matter generates small particles and waves, creating dense formations. These particles, dense and possessing weight, contribute to forming solid objects in the universe.

Subatomic Black Holes

Particles can be seen as subatomic-black holes, made up of tightly packed electrons too small to form actual black holes. These electrons remain in close, tight formations, contributing to the density and mass of particles.

By reconciling these ideas, we gain a deeper understanding of the universe's expansion, the nature of time, and the structure of reality. This exploration reveals the complex interplay between fundamental particles, time, and the ever-expanding cosmos.

Chapter 47.6: Reconstructing the Universe's Geometry

Understanding the Basic Building Block

The fundamental building block of objects in the universe starts with electrons. According to the laws of physics, the force of impact increases with the square of the increase in speed. Pressure, defined as stress, is a scalar quantity given by the magnitude of force per unit area. In gases, this force is exerted by the change of momentum of molecules impinging on surfaces, resulting in pressure acting on solids.

The Role of Electrons

Electrons, in tight formation and moving at near light speed, create the illusion of weight due to their inherent properties as particles. To recreate the universe's geometry, we begin with its primary building blocks: electrons. We must determine if the universe is constructed from an illusion field or substantive matter composed of electrons.

The Protective Field of the Universe

Assuming the universe exists and protects us from the void, everything within it comprises electrons, protons, and neutrons. Electrons, as fundamental particles, serve as the first dimension we encounter. Their actual shape and dimensional level remain unknown, but we surmise that electrons and atoms are multidimensional. This concept forms the first level of dimension we encounter.

The Concept of an Illusion Field

Imagine the universe created based on the illusion field. The idea that we are walking particles held together by various forces, composed of trillions of atoms, is a fascinating thought. Within this field, the universe's pulsating waves collapse and expand, designing and sculpting the objects we know today. Each entity is formed by various three-dimensional shapes.

Dimensions and Chemical Elements

1. First Dimension (Subatomic Level): Smaller than an atom, these are the fundamental building blocks of matter.

2. Second Dimension (Atomic Level): Atoms, which can be found in the periodic table of chemical elements.

3. Third Dimension (Molecular Level): Molecules, groups of atoms bonded together, representing the smallest unit of a chemical compound that can take part in chemical reactions.

4. Fourth Dimension (Chemical Compounds): Unique substances combining two or more elements in fixed proportions, always maintaining the same composition. The smallest particle of most compounds in living things is a molecule. The outer layer of an object determines its chemical nature, composed of more than one element chemically bonded.

Homogeneous and Heterogeneous Compounds

Chemical compounds can be either homogeneous or heterogeneous. Homogeneous compounds consist of similar elements throughout, while heterogeneous mixtures have non-uniform compositions and consist of multiple phases. For example, oil and water form two separate layers, representing a heterogeneous mixture.

Multi-Dimensional Shapes

Objects can be composed of various three-dimensional shapes, encompassed by a larger dimension of 3D shapes. A rock, for instance, consists of multiple dimensions, with the rock itself being a higher summation of all-encompassing three-dimensional shapes within it.

By understanding the basic building blocks and the multi-dimensional nature of objects, we gain insights into the complexity and beauty of the universe. This exploration reveals how fundamental particles, dimensions, and chemical compounds come together to form the reality we perceive.

Chapter 48.0: Geometry - The Key to Understanding the Universe

The Significance of Geometry

Geometry is fundamental to understanding the universe. The way shapes and figures scale, and how heat dissipates from them, offers insights into the conditions of the universe, whether it was hot or cold before the Big Bang.

Heat Dissipation and the Geometry of the Universe

Consider a hypothetical 1" x 1" x 1" cube of Kryptonite, producing one joule of energy per second. If this cube loses 0.1 joules of energy for each square centimeter of its surface area exposed to air, and it has six faces, it will lose a total of 0.6 joules of energy per second.

The Principles of Scaling in Geometry

In geometry, scaling a figure by a factor of K leads to the following changes:

- All lengths change by a factor of K.
- All areas change by a factor of K^2.
- All volumes change by a factor of K^3.
- All angles remain the same.

These principles reveal that as objects scale up, their volumes increase at a more rapid rate than their surface areas.

Understanding Pre-Big Bang Conditions

By applying these geometric principles, we can infer conditions before the Big Bang. Heat dissipation from objects in space can help us understand whether the early universe was hot or cold. The energy loss from objects due to their surface areas plays a crucial role in these determinations.

Applying Geometric Scaling

Let's expand on the example of the Kryptonite cube:

1. Initial Size: A 1" x 1" x 1" cube.
   o Surface Area: 6 cm6^2
   o Volume: 1 cm1^3
   o Energy Loss per Second: 0.6 joules
2. Scaling by a Factor of 2:
   o New Length: 2"
   o New Surface Area: $6 \times 4 = 24\ cm^2$
   o New Volume: $1 \times 8 = 8\ cm^3$
   o Energy Loss per Second: $24 \times 0.1 = 2.4$ joules
3. Scaling by a Factor of 3:
   o New Length: 3"
   o New Surface Area: $6 \times 9 = 54\ cm^2$
   o New Volume: $1 \times 27 = 27\ cm^3$
   o Energy Loss per Second: $54 \times 0.1 = 5.4$ joules

This demonstrates that as the cube scales, its volume increases more rapidly than its surface area, affecting the rate of energy loss.

Conclusion

Scaling in geometry illustrates that volumes increase at a more rapid rate than areas. This principle helps us understand the energy dynamics in the universe, contributing to our knowledge of pre-Big Bang conditions and the universe's thermal history.

By delving into these geometric principles, we can gain deeper insights into the fundamental nature of the universe and its behavior over time.

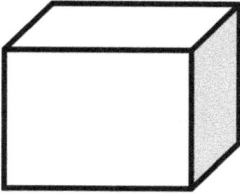

1x1x1
1 Joule
0.6 joules
0.6/1
60%

Chapter 48.1: Understanding Energy Dynamics in Geometric Scaling

Exploring Cube Dimensions and Energy Dynamics

To understand the relationship between geometry and energy dynamics, let's delve into the scaling of cubes and their associated energy properties. We will examine how the size of a cube affects the energy produced and lost per second, and the resulting ratio of energy lost to energy produced.

Cube Dimensions and Energy Calculations

Let's consider three different cube dimensions and analyze the energy dynamics for each:

Cube Dimension (in cm)	Energy Produced per Second (joules)	Energy Loss per Second (joules)	Ratio (Energy Lost/Energy Produced)
2 x 2 x 2	8	2.4	30%
5 x 5 x 5	125	15	12%
10 x 10 x 10	1000	60	6%

Detailed Calculations

1. **Cube Dimension: 2 x 2 x 2 cm**

   o   Energy Produced per Second: 8 joules

   o   Surface Area: 6 faces x (2 x 2) = 24 cm^2

   o   Energy Loss per Second: 24 cm^2 x 0.1 joules/cm^2 = 2.4 joules

   o   Ratio: 2.4 / 8 = 30%

2. **Cube Dimension: 5 x 5 x 5 cm**

   o   Energy Produced per Second: 125 joules

   o   Surface Area: 6 faces x (5 x 5) = 150 cm^2

   o   Energy Loss per Second: 150 cm^2 x 0.1 joules/cm^2 = 15 joules

   o   Ratio: 15 / 125 = 12%

3. **Cube Dimension: 10 x 10 x 10 cm**

   o   Energy Produced per Second: 1000 joules

   o   Surface Area: 6 faces x (10 x 10) = 600 cm^2

   o   Energy Loss per Second: 600 cm^2 x 0.1 joules/cm^2 = 60 joules

   o   Ratio: 60 / 1000 = 6%

Insights from Scaling and Energy Dynamics

From these calculations, we observe that as the cube's dimensions increase, the ratio of energy lost to energy produced decreases. This demonstrates that larger volumes have a relatively smaller surface area compared to their volume, leading to less energy loss per unit of energy produced.

Conclusion

The study of geometric scaling and energy dynamics highlights how the properties of shapes influence energy efficiency. Larger objects tend to lose a smaller percentage of their produced energy, making them more efficient in terms of energy retention.

Understanding these principles not only provides insights into geometric scaling but also sheds light on the fundamental nature of energy interactions in physical objects.

Chapter 48.2: The Relationship Between Energy Loss and Volume

Understanding Energy Loss and Volume Increase

Energy loss decreases in size as volume increases. This phenomenon is rooted in the principles of geometric scaling. When the size of an object increases, its volume changes by a factor of k^3, while its surface area changes by a factor of k^2. As a result, the energy produced by the object grows faster than the energy lost.

Scaling Laws and Energy Dynamics

To grasp this concept, let's delve into the scaling laws that govern geometric shapes:

- Volume (V) changes by a factor of k^3.
- Surface Area (A) changes by a factor of k^2.

As an object's size increases, its volume increases at a much faster rate than its surface area. This means that larger objects have a relatively smaller surface area compared to their volume, leading to a slower rate of energy loss per unit of energy produced.

The Slow Dissipation of Energy in the Universe

The universe dissipates energy very slowly due to this geometric principle. At the beginning of the universe, following the Big Bang, the universe was extremely hot. As the universe expanded, its volume increased significantly, causing the temperature to decrease over time. This gradual cooling is a direct consequence of the relationship between volume and surface area.

Practical Example: Energy Dynamics in a Cube

To illustrate this concept, let's revisit our example of cubes with different dimensions:

1. **Cube Dimension: 2 x 2 x 2 cm**
   - Volume (V): $2^3 = 8$ cm^3
   - Surface Area (A): $6 \times (2 \times 2) = 24$ cm^2
   - Energy Loss: $24 \times 0.1 = 2.424$ joules
2. **Cube Dimension: 5 x 5 x 5 cm**
   - Volume (V): $5^3 = 125$ cm^3
   - Surface Area (A): $6 \times (5 \times 5) = 150$ cm^2
   - Energy Loss: $150 \times 0.1 = 15$ joules
3. **Cube Dimension: 10 x 10 x 10 cm**
   - Volume (V): $10^3 = 1000$ cm^3
   - Surface Area (A): $6 \times (10 \times 10) = 600$ cm^2
   - Energy Loss: $600 \times 0.1 = 60$ joules

As seen from these calculations, the ratio of energy loss to energy produced decreases as the cube's dimensions increase.

Implications for the Universe's Temperature

In the early universe, the high temperature resulted from the small initial volume after the Big Bang. As the universe expanded and its volume increased, the temperature decreased due to the slower rate of energy dissipation. This gradual cooling over billions of years allowed the formation of galaxies, stars, and planets.

By understanding the principles of geometric scaling and energy dynamics, we gain insights into the behavior of the universe from its hot, dense beginnings to its current state. This knowledge helps us appreciate the intricate interplay between volume, surface area, and energy loss.

Chapter 49.0: Redefining Geometry

Changing the Perspective on Geometry

We can reimagine geometry in fascinating ways. It's intriguing to think that not everything is as it seems, and even geometry can have flexible laws. Having taken an advanced online course in geometry, my insatiable appetite for this subject led me to explore various geometric behaviors, including hyperbolic geometry, finite geometry, and even the whimsical Taxi-Cab geometry.

Fundamental Assumptions in Geometry

The first fundamental assumption in traditional Euclidean geometry is that the sum of all angles in a triangle equals 180 degrees. Another basic principle is the similarity of triangles, characterized by postulates such as:

- Side-Angle-Side (SAS): Two triangles are similar if two sides and the included angle of one are proportional to the corresponding sides and angle of the other.
- Side-Side-Side (SSS): Two triangles are similar if all three sides of one are proportional to the three sides of the other.
- Angle-Angle-Angle (AAA): Two triangles are similar if all three angles of one are equal to the corresponding angles of the other.

However, these principles do not always hold true in spherical geometry.

Exploring Spherical Geometry

In spherical geometry, the differences between familiar shapes can be striking. For example:

- A Polygon is a plane figure bounded by straight edges.
- A Bigon (rarely used) is a polygon with two edges and two vertices.

A Thought Experiment: The Woman's Walk

Consider a woman walking on the spherical surface of Earth. As she walks, she makes a series of three 90-degree turns. In traditional Euclidean geometry, this would create a 270-degree turn, leading her to face a new direction. But on a spherical surface, something different happens. After three 90-degree turns, she ends up back at her starting point, having completed a triangular journey on the globe.

This thought experiment highlights the fascinating deviations of spherical geometry from traditional Euclidean concepts.

The Flexibility of Geometric Laws

Exploring these different geometries demonstrates the flexibility of geometric laws. Each type of geometry—whether Euclidean, spherical, hyperbolic, finite, or even the playful Taxi-Cab geometry—offers unique insights and challenges our conventional understanding.

By redefining and reimagining geometry, we gain a deeper appreciation for the diverse and dynamic nature of mathematical principles. Whether it's navigating the curved surface of a sphere or contemplating the abstract spaces of hyperbolic geometry, each perspective enriches our knowledge and understanding of the world around us.

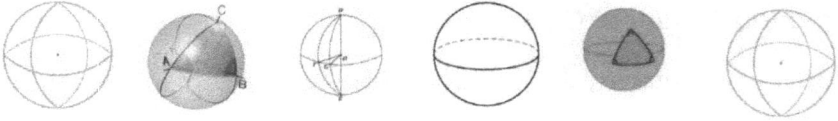

If possible, rewrite and expand on the topic. Start with Chapter 49.1."Spherical geometry is the geometry of the two-dimensional surface of a sphere. In this context, the word "sphere" refers only to the 2-dimensional surface, and other terms like "ball" or "solid sphere" are used for the surface together with its 3-dimensional interior. There is no such thing as a bigon being a two–side polygon in our flat geometry in spherical geometry. Area in Spherical geometry A = \frac{x}{360} x total surface area. A Spherical triangle drawn on the sphere's surface consists of three edges for the second above image. <A is part of one of the bigon-shaded areas in brown. Bigon starts at "A" comedown and disappears into the southern hemisphere in green, surrounded by two bigon on each side. They are also bigon for B and C. Area = \frac{A}{360} x total surface area Area = \frac{B}{360} x total surface area Area = \frac{C}{360} x total surface area Notice: all other bigons in the first picture above cover the center bigon. The highlighted area covers three times the sum of these three bigons and is the whole Northern Hemisphere, half the sphere, plus extra covering of the center triangle. Sum it up: The sum of those three bigons is the whole Northern Hemisphere, half the sphere, plus the extra coverings of the center triangles. Formula: ½ total surface area + 2x area triangle = \frac{\left(A+B+C\right)}{360} x total surface area. Half the total surface area of the sphere is the Northern Hemisphere, plus the two extra coverings of a given triangle must add up to the areas of the three bigons \frac{A}{360} + \frac{B}{360}+ \frac{C}{360} x the total surface

area. Simplify: There are lots of twos and halves. Half the total surface area on the left, some fractions of the total surface area on the right, bring those terms together and divide everything in two. Simplify the formula: Area of triangle = \frac{\left(A+B+C-180\right)}{720} x total surface area Example revisiting our woman walk on the sphere. Area of triangle = \frac{\left(90+90+90-180\right)}{720} x total surface area = \frac{1}{8} x total surface area. Lady walk makes a 90° turn to come back to where she started. The formula suggests that the center triangle in the picture is one-eight of the total surface area. That angle at the top is one-quarter of that 360 at the top \frac{90}{360} = \frac{1}{4} Northern Hemisphere. The Northern Hemisphere is half the sphere. We have one-quarter of one-half the sphere. That is \frac{1}{8} of the surface area of the sphere—the formula work.

Chapter 49.1: Spherical Geometry - A Different Perspective on Shapes

Understanding Spherical Geometry

Spherical geometry is the study of the two-dimensional surface of a sphere. In this context, the term "sphere" refers specifically to the 2-dimensional surface, whereas terms like "ball" or "solid sphere" include both the surface and its 3-dimensional interior.

The Concept of a Bigon in Spherical Geometry

Unlike flat geometry, spherical geometry does not accommodate the concept of a bigon, a two-sided polygon. Instead, the area calculations in spherical geometry differ significantly from those in Euclidean geometry.

Calculating Areas in Spherical Geometry

The area of a spherical shape can be calculated using the formula:

$$\text{Area} = \frac{x}{360} \times \text{Total Surface Area}$$

For instance, consider a spherical triangle drawn on the surface of a sphere. This triangle consists of three edges, and its area is part of the bigon-shaded regions.

Example: Bigon Calculation

Let's examine a bigon starting at point "A" and extending into the southern hemisphere. This area is surrounded by other bigons at points B and C.

$$\text{Area=A} = \frac{A}{360} \times \text{Total Surface Area}$$

$$\text{Area=B} = \frac{B}{360} \times \text{Total Surface Area}$$

$$\text{Area=C} = \frac{c}{360} \times \text{Total Surface Area}$$

Notice that the highlighted area covers three times the sum of these three bigons, equating to the entire Northern Hemisphere (half the sphere), plus an extra covering of the center triangle.

Summing Up the Areas

The combined area of the three bigons covers the whole Northern Hemisphere, half the sphere, plus the additional area of the center triangle. This can be represented by the formula:

$\frac{1}{2}$ Total Surface Area + 2 × Area of Triangle = $\frac{A+B+C}{360}$ x Total Surface Area

By simplifying, we recognize that the Northern Hemisphere plus the extra area of a given triangle must equal the sum of the areas of the three bigons:

$$\frac{A}{360} + \frac{B}{360} + \frac{C}{360} \times \text{Total Surface Area}$$

Combining and dividing these terms, we derive:

Area of Triangle $= \frac{A+B+C-180}{720} \times$ Total Surface Area

Revisiting the Woman's Walk

Consider the example of a woman walking on the sphere. If she makes three 90° turns, she returns to her starting point. Using our formula, the area of the triangle formed by her path is:

$$\text{Area of Triangle} = \frac{90+90+90-180}{720} \times \text{Total Surface Area} = \frac{1}{8} \times \text{Total Surface Area}$$

This calculation suggests that the triangle formed by her walk is one-eighth of the sphere's total surface area. The angle at the top is one-quarter of 360 degrees:

$$\frac{90}{360} = \frac{1}{4} \text{Northern Hemisphere}$$

Since the Northern Hemisphere is half the sphere, we have:

$$\frac{1}{4} \times \frac{1}{2} = \frac{1}{8}$$

Thus, the formula works, confirming that the area of the triangle is one-eighth of the sphere's total surface area.

By exploring spherical geometry, we gain a different perspective on shapes, angles, and areas, challenging our conventional understanding and broadening our knowledge of mathematical principles.

Chapter 50.0: The Creation of Physical Space

How Was Physical Space Created?

When pondering the universe, we often wonder about the nature and creation of physical space. How did it come into existence, and what is its working definition? Physical space can be understood as a realm where two objects are separated by time and motion. Though we might not move, the physical space between objects can increase or decrease depending on the forces exerted.

The Big Bang and the Creation of the Universe

To comprehend the creation of the universe, let's begin with the concept of "nothing." When I say "nothing," I mean the complete absence of everything—no lines, no thoughts, no logical absolutes, nothing real or imaginary. Before the Big Bang, there were only two logical possibilities: nothing or something. Initially, there was nothing, and then something happened, bringing reality into existence.

It doesn't matter whether there were multiple entities or just one; they would all have existed as one. This flip-flop between nothing and something illustrates that once something exists, the concept of nothing ceases to be. This challenges the multi-multiverse theory, which suggests multiple "somethings" emerging from nothing. It's a binary state: either there is something, or there is nothing. If something exists, then everything exists within that existence.

The Nature of Existence

Within this existence, there is no shape, as there is no space for anything to form a shape. However, inside that existence, there is motion and time. Therefore, within that existence, space also exists. We assume that motion is linear. Just before the Big Bang, at $T = -0$, we can refer to the Cartesian coordinate system to visualize this concept.

Visualizing the Initial State

Imagine a point in the Cartesian coordinate system, representing the initial state of nothingness. This point signifies the absence of space and time. As the Big Bang occurs, this point expands, creating physical space and time. The Cartesian coordinates help us understand the expansion of space from a single, dimensionless point to a vast, multidimensional universe.

The Role of Geometry in Understanding Space

Geometry plays a crucial role in understanding the creation of the universe. By using simple geometric principles, we can visualize the expansion of space and the separation of objects by time and motion. The initial expansion of the universe created space, within which objects could move and interact.

Conclusion

The creation of physical space is a profound concept rooted in the transition from nothing to something. The Big Bang marks the beginning of this transition, where a dimensionless point expanded to form the universe, we know today. By exploring these ideas through geometry and the Cartesian coordinate system, we gain a deeper understanding of the fundamental nature of existence and the creation of physical space.

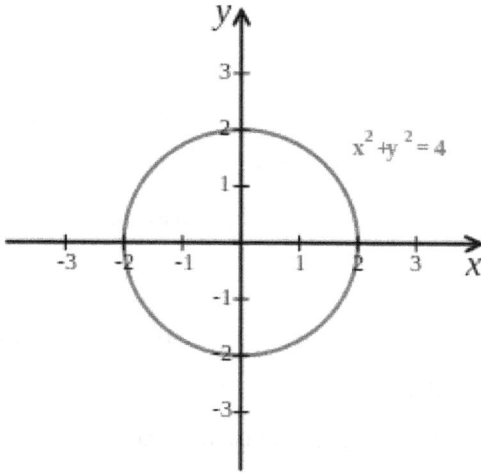

The graph shows a circle centered at the origin with equation $x^2 + y^2 = 4$ on a Cartesian coordinate system, with the x-axis and y-axis marked from -3 to 3.

Chapter 50.1: The Creation of Space in a New Existence

From Nothing to Something

Now that nothing no longer exists in the universe, something has come into being, occupying space. To understand how space is created and occupied, let's delve deeper into the concept of time and movement within this new existence. Here, we'll slow time to observe how something begins to take up space.

Defining Movement and Vibration

In this new existence, movement can be defined as a form of vibration. By slowing the time before the Big Bang explosion, we can examine the happenings within this existence. Imagine starting at a fixed point on a Cartesian coordinate system's horizontal line at 0°. All movement within this new reality is linear and occurs in all directions.

The Formation of Dimensions

As linear movements fill the quadrant and reach a ninety-degree angle, they create a two-dimensional object, forming an area. Initially, these movements were mere lines with a dimensional scaling factor of k^1. At ninety degrees, they transform into an "Area" with a scaling factor of k^2.

As this reality progresses towards 180°, another phenomenon occurs, giving rise to a three-dimensional space (k^3) by adding "volume"— another dimension to space. This process continues, and upon reaching the -270° angle, yet another dimension is added, leading to the creation of a fourth dimension (k^4), which we can refer to as a "Tesseract."

This process continues until K^{360}. In a circle, there are 360 vectors, each capable of creating an object. The combination of various instances of reality forms a structure. This is why each instant of existence is multi-dimensional, combining to create diverse shapes and dimensions.

Understanding the Creation of Space

Starting at a fixed point, various rays extend outward from this position. The space between these rays is nothingness, or what we can call "space." Motion is required to cross the gap between these rays, as the opening represents nothingness. This motion creates an imaginary distance between the rays, known as the newly developed area dimension.

Expanding Dimensions and Space

The creation of space is a dynamic process. As more dimensions are created, more space is formed, and the area increases. Each new dimension adds a layer of complexity and depth to the fabric of the universe, expanding the boundaries of space and creating new realms of existence.

By understanding these principles, we gain insights into the intricate process of how space is formed and occupied within the universe. The transition from nothing to something marks the beginning of a fascinating journey through the dimensions of reality.

Chapter 50.2: Geometry Behind the Expansion of Space

The Geometry of Expansion

As the universe expands, new space is created, and the area increases due to the scaling effect of each dimension expanding faster than the previous one. Let's delve into the geometry behind this process. We begin with a line (K^1), then an area (K^2), followed by volume (K^3), and finally a tesseract (K^4). This process repeats, creating layers upon layers of dimensions until it reaches K^{360} and starts anew.

Layered Construction of the Universe

The construction of the universe involves the stacking of dimensions, with volume increasing faster than area, and tesseracts increasing faster than volume. Each new dimensional creation forces the area to expand to accommodate the new dimension, resulting in an incremental scaling by a factor of K^x. This expansion occurs in a factorial manner, represented mathematically by the symbol (!), where n! multiplies n by every preceding number.

The Big Bang and the Mushroom Cloud Effect

The Big Bang can be visualized like a mushroom cloud, developing or expanding similarly to a magnetic field. The growth doesn't push outward in a linear fashion. Instead, the lower dimension starts at the center and expands over the preexisting layer. Each new dimension is created on top of the previous one, causing the area to grow with the new attribute.

Visualizing the Universe

From a top or bottom view, the universe appears as a round plate with various dimensional rings, similar to the annual growth rings of a tree stump. When viewed sideways, the universe resembles a folding telescope with different cylindrical stages, ranging from large to small. The smaller circle represents the line or first dimension, while the largest circle represents the last dimension or stage.

Linear time moves in the direction of the larger cylindrical circle. Observing the universe from the top, the smaller circle is at the center. We exist within the third circle. The first circle is the first dimension, the second circle is the second dimension, and the third circle is the third dimension. Viewing the universe sideways, you can see the structure of various geometric shapes corresponding to each ring.

The Scaling Effect

As dimensions increase, the area and volume also increase, creating more space. The process involves:

- Starting at a fixed point, with various rays extending outward.
- The space between the rays represents nothingness, which is crossed by motion.
- This motion creates an imaginary distance between the rays, known as the newly developed area dimension.
- More dimensions lead to more space, increasing the area incrementally.

By understanding these geometric principles, we gain insights into how space is created and expanded within the universe. The layered construction and scaling effect illustrate the complex nature of dimensional expansion and its impact on the cosmos.

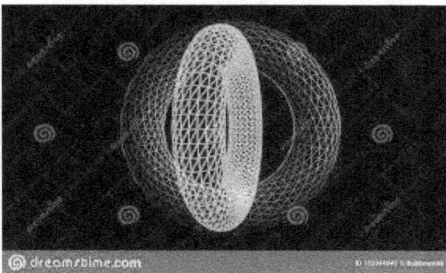

Chapter 50.3: The Fabric of Space and the Nature of Time

Time as the Fabric of Space

Time is not merely a dimension but the very fabric of space itself. Time and motion collectively create space. As we move across the universe, the distance between two objects diminishes, effectively reducing and contracting space. To better understand this, let's reexamine Einstein's theory.

Einstein's Theory of Relativity

According to Einstein's theory of relativity, the universe imposes a speed limit— the speed of light. This limit is mathematically provable. Let's start with two distant points. To travel between these points, velocity is required. Based on observations, as velocity increases, the time taken to travel between the two points decreases.

The Concept of Instantaneous Travel

There comes a point where velocity increases so much that the distance between the two points becomes almost instantaneous, approaching time zero. However, according to Einstein, time can never actually reach zero because there is a speed limitation before time equals zero.

Challenging Einstein's Perspective

Suppose, for the sake of argument, that Einstein was wrong. If it were possible to reach time zero, what implications would this have on our understanding of space and time?

Time and Motion: If time and motion create space, then reaching time zero would mean collapsing the fabric of space. This suggests that space could be infinitely compressed or expanded.

Instantaneous Travel: Achieving time zero would imply instantaneous travel across any distance, effectively nullifying the concept of space as we know it.

Dimensional Impact: The collapse or expansion of space-time would have profound effects on the dimensional structure of the universe, potentially altering the very nature of reality.

Visualizing Space-Time

To visualize this, imagine a fabric stretched out, representing space-time. As an object moves across this fabric, it creates ripples, representing motion and the passage of time. If the object moves faster, the ripples become more pronounced, reducing the perceived distance. If the object could move infinitely fast, the fabric would fold upon itself, making distant points touch.

Implications of Time Zero

Reaching time zero could have several implications:

- **Space Compression:** Space could be compressed to a singular point, collapsing the universe into a single location.
- **Infinite Expansion:** Conversely, space could expand infinitely, making all points equidistant.
- **Altered Reality:** The fundamental nature of reality would be altered, challenging our current understanding of physics and the universe.

Chapter 50.4: The Illusion of Motion in Space-Time

Time Zero and the Nature of Travel

By exploring these concepts, we gain deeper insights into the relationship between time, motion, and space. Whether Einstein's theory holds or not, the fabric of space-time continues to captivate and challenge our understanding. When we speak of "time zero," we must ask: have we arrived at our destination instantaneously, or have we propelled ourselves into oblivion?

Einstein's Infinite Resistance

In Einstein's equations, infinite resistance equates to oblivion or nothingness, effectively merging the two. This resistance is not due to actual movement, as we exist within a three-dimensional field. Instead, collapsing and expanding space-time between two points creates the illusion of motion. The Big Bang, in this context, can be seen as a dimensional explosion, growing from a fixed position rather than being drawn out.

The Concept of K^0

According to my proposed theory, there is also K^0. At K^0, there is no dimension, or it acts as a void that prevents runaway time. In this expanding universe scenario, runaway time would lead to infinity. This can be compared to a computer game, where a character does not move; instead, the field adjusts to simulate movement.

Simulating Movement in a Computer Game

In a computer game, the Cartesian coordinate system diminishes space between characters by adding or subtracting coordinates. This manipulation allows the mind to engage with the earth's coordinate system, maneuvering characters seamlessly. Forces are either added or subtracted from the environment to facilitate movement. Sense detects a dense particle field to determine whether an object is in the way. Essentially, the species recreates the universe's motion by manipulating space and time for movement or activity.

The Role of Cartesian Coordinates

Using Cartesian coordinates, we observe how movement and time interact to create space. By adjusting coordinates, the system simulates motion, echoing the principles of space-time manipulation. This simulation illustrates how collapsing and expanding space-time can give the illusion of movement, mirroring the universe's behavior.

The Structure of the Universe

The universe is not static; it grows from a fixed position, expanding and creating new dimensions. This expansion can be visualized as layered growth, with each layer representing a new dimension. As time moves away from the Big Bang, the universe continues to expand, creating more space and dimensions.

By understanding these principles, we gain a clearer picture of how space-time functions and the nature of movement within it. Whether through Einstein's equations or through theoretical models, the exploration of space-time remains a profound and captivating endeavor.

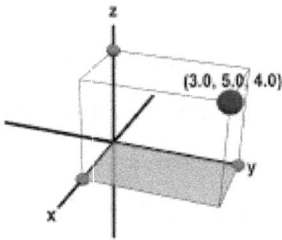

Chapter 50.5: The Dynamics of Time and the Expanding Universe

The Concept of Runaway Time

The expanding universe theory operates on the concept of linearly polarized time following the Big Bang explosion, creating what can be described as runaway time. In this theory, the universe will continue to expand from the moment of the Big Bang toward infinity, only to potentially restart at some point in the future. This concept envisions time moving in a linear fashion, stretching endlessly as the universe grows.

Time Collapsing and Expanding

Contrasting the expanding universe theory, another perspective suggests that time could eventually collapse and retract like an accordion. When examining a telescope drawing, one can see that the final dimensional loop could trigger another growth phase for the universe. In this model, the force of creation inherent in the last dimensional loop appears simultaneously with the initial time rings as they close. This simultaneous appearance pushes the time ring forward, propelling the universe into a new phase of expansion.

Visualizing the Universe's Structure

This dynamic process would result in a universe that resembles a spherical structure composed of various layers, each filled with multiple objects. Each layer represents a different dimensional phase, contributing to the overall complexity and richness of the universe.

1. **Layered Expansion:** The universe expands in layers, with each new dimension adding to the existing structure. This layering process creates a multi-dimensional, spherical universe.

2. **Dimensional Interplay:** The interplay between dimensions and the forces of creation and collapse contributes to the universe's dynamic nature. As one dimensional loop closes, it triggers the expansion of the next, creating a continuous cycle of growth.

3. **Visual Representation:** From a visual standpoint, the universe would appear as a series of concentric spheres or rings, each representing a different dimension and phase of expansion. The innermost rings would represent the earliest phases, while the outermost rings correspond to the most recent expansions.

Implications of Dimensional Forces

The concept of dimensional forces driving the expansion and collapse of the universe has several intriguing implications:

- **Continuous Evolution:** The universe is in a state of continuous evolution, constantly expanding and collapsing in cycles.
- **Multi-Dimensional Interactions:** The interactions between various dimensions and the forces of creation and collapse add complexity to the universe's structure.
- **Spherical Universe:** The spherical structure of the universe, with its layered composition, highlights the intricate interplay between dimensions.

By exploring these concepts, we gain a deeper understanding of the universe's dynamic nature and the role of time and motion in shaping its structure. Whether viewed through the lens of the expanding universe theory or the accordion-like collapse and expansion model, the study of space-time continues to reveal captivating insights into the nature of our cosmos.

Chapter 51.0: Reducing God into a Geometric Mathematical Equation

Understanding the Basics of Geometry

A point is a 0-dimensional mathematical object specified in n-dimensional space using an n-tuple of coordinates. In dimensions greater than or equal to two, points are sometimes considered synonymous with vectors, so points in n-dimensional space are sometimes called n-vectors. In geometry, a point is defined as a location with no size. It has no width, length, or depth.

Fundamental Geometric Concepts

1. Point: A locatioen with no size, and zero dimensions.
2. Line: Extends infinitely in either direction, one-dimensional with no width.
3. Plane: Extends infinitely in two dimensions.
4. Euclidean Space: Virtual space of geometry representing physical space, including three-dimensional space and the Euclidean plane.

Scientists used the qualifier "Euclidean" to distinguish Euclidean spaces from other spaces considered in modern mathematics and physics.

Points in Euclidean Geometry

A mathematical point has zero dimensions; it has no length, area, or volume. Points are fundamental building blocks of geometry, and more complex spaces and structures are made from uncountably many points related to each other. Points are undefined terms that don't require a formal definition, and they are labeled with one capital letter.

Lines and Planes

A line is an infinitely long straight mark or band, while a plane is an infinitely extending two-dimensional surface. In Euclidean space, these concepts form the foundation of geometric constructs.

God and Geometry

Psalm 90:2 states, "From forever in the past to forever in the future, you are God." This God is unlimited in power, beyond time and space constraints. He created a universe with no beginning and end (Genesis 1:1).

His name Jehovah means He who was in the past, who presently is, and who will be in the future. He said, "I am the Alpha and Omega, the Beginning and the End, the First and the Last, and behold, I am alive always" (Revelation 1:8 & 11; 2:6; 22:18). He also said, "Before Abraham was, I am" (John 8:56–58), and "My Father worked in the past, and I am working now" (John 5:17–18).

Biblical References

- John 1:1–2: "At the beginning (a timeless beginning) was the Word (the Lord Jesus, God's Son), and the Word was with God, for the Word is God. The same was in the beginning with God."
- Genesis 1:1: "At the beginning (of time) God created the Heavens and the Earth."
- Isaiah 9:6: "For unto us a child is born, unto us a Son is given, and the government shall be upon His shoulders: and His Name shall be called Wonderful, Counselor, The Mighty God, The Father of Eternity, The Prince of Peace."

- John 3:13: "No man (person) has ascended to Heaven, but He that came down from Heaven, even the Son of Man Who is in Heaven."

- Philippians 2:6, 10-11: "God created everything in heaven and on earth, visible and invisible, whether thrones or dominations, principalities or powers; all were created through Him and for Him."

- Proverbs 15:3: "The eyes of the Lord are in every place, keeping watch on the evil and the good."

- Acts 17:24 ESV: "The God who made the world and everything in it, being Lord of heaven and earth, does not live in temples made by man."

- Jeremiah 23:23-24: "Am I a God at hand, declares the Lord, and not a God far away? Can a man hide in secret places so I cannot see him? Declares the Lord. Do I not fill heaven and earth? Declares the Lord."

God's name Jehovah means He who was in the past, who presently is, and who will be in the future. This powerful being is described as eternal, omnipotent, and omnipresent, transcending all dimensions of time and space.

Conclusion

By understanding geometry, we can appreciate the fundamental building blocks of the universe. Points, lines, and planes form the basis of geometric constructs, which, when combined, create complex structures and spaces. Similarly, the concept of God transcends time and space, embodying eternal existence and omnipotence.

Euclidean space

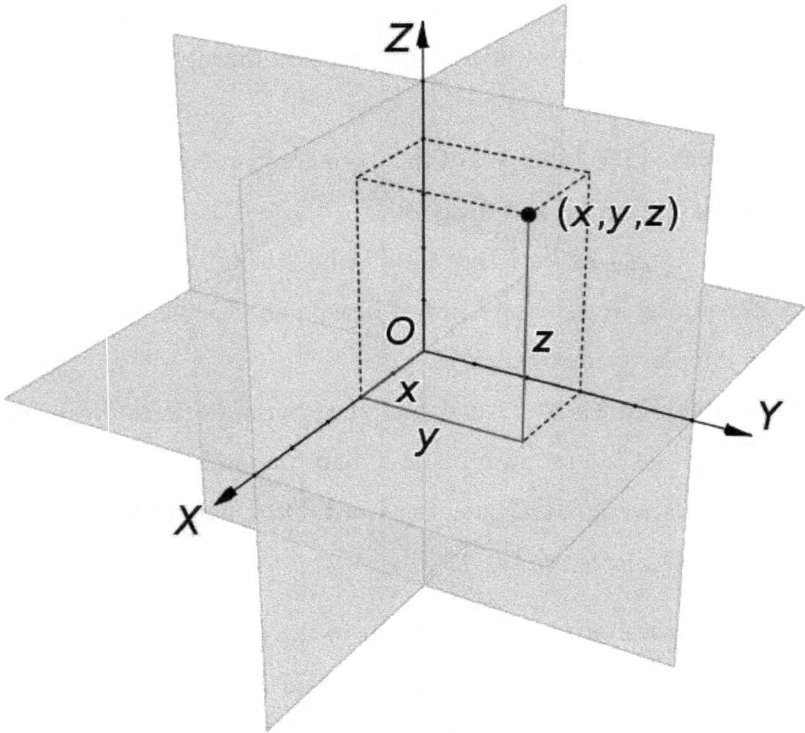

Chapter 51.1: The Alpha and the Omega

The Eternal Nature of God

It becomes evident that God is the Alpha and the Omega, which means the beginning and the end. According to the Bible, God has no beginning or end and exists in everything. From this, one can deduce that God has no dimension in space and time. God is composed of infinite parts, existing in a set location in Euclidean space.

Geometric Analogy

In geometry, a point has no dimension; it lacks height, width, and depth, thereby possessing zero dimensions. A point at which a function takes an infinite value, especially in space-time, can be defined here as an instant of reality. Is it possible to conceptualize God mathematically by considering time?

Biblical References and Time

The Bible describes the creation of the heavens, the earth, and time itself. The one who can accurately calculate time and predict it controls the universe. This probability matrix suggests that time continually happens like runaway time. God, as described in the Bible, has no beginning or end, akin to a point with no dimensions.

The Beginning of Creation

"In the beginning, God created the heavens and the earth. The earth was formless and empty, darkness was over the surface of the deep, and the Spirit of God hovered over the waters. God said, 'Let there be light,' and there was light" (Genesis 1 NIV). This narrative, though simplified, illustrates the creation of something tangible from nothing. Time did not exist initially, leaving no measure for the duration of this appearance or disappearance.

Time and Reality

In the beginning, there was zero time. The function that takes an infinite value in space-time laid the foundation for constructing a probability universe. This matrix, set by God, encompasses all possible outcomes. When an instant of reality moves from one location to another, it creates time. This movement, like the concept in "The Twilight Zone," lies between light and shadow, science and superstition.

God's Eternal Presence

God declared, "I am the Alpha and Omega, the Beginning and the End, the First and the Last, and behold, I am alive always" (Revelation 1:8 & 11; 2:6; 22:18). "Before Abraham was, I am" (John 8:56–58). "My Father worked in the past, and I am working now" (John 5:17–18). "At the beginning (a timeless beginning) was the Word (the Lord Jesus, God's Son), and the Word was with God, for the Word is God" (John 1:1–2).

God and the Creation

"In the beginning (of time), God created the heavens and the earth" (Genesis 1:1). Isaiah 9:6 says, "For unto us a child is born, unto us a Son is given, and the government shall be upon His shoulders: and His Name shall be called Wonderful, Counselor, The Mighty God, The Father of Eternity, The Prince of Peace." John 3:13 notes, "No person has ascended to Heaven, but He that came down from Heaven, even the Son of Man Who is in Heaven."

The Omnipresence of God

God created everything in heaven and on earth, visible and invisible, whether thrones or dominions, principalities or powers; all were created through Him and for Him (Philippians 2:6, 10-11). Proverbs 15:3 states, "The eyes of the Lord are in every place, keeping watch on the evil and the good." Acts 17:24 ESV declares, "The God who made the world and everything in it, being Lord of heaven and earth, does not live in temples made by man." Jeremiah 23:23-24 asks, "Am I a God at hand, declares the Lord, and not a God far away? Can a man hide in secret places so I cannot see him? Declares the Lord. Do I not fill heaven and earth? Declares the Lord."

Conclusion

By understanding geometry, we gain insights into the fundamental building blocks of the universe. Points, lines, and planes form the basis of geometric constructs, which combine to create complex structures and spaces. Similarly, the concept of God transcends time and space, embodying eternal existence and omnipotence.

Chapter 51.2: The Interface Between Nothingness and Reality

The Beginning and the End

We have a beginning and an end as the instant of reality jumps instantaneously from one point to another. Drawing a line in the nothingness points to an imaginary beginning—the point of entry into the nothingness. Since nothingness doesn't exist and the start is just an imaginary beginning, the singularity is also imaginary.

Understanding Singularity

A gravitational singularity, spacetime singularity, or simply singularity is a condition in which gravity is so intense that spacetime itself breaks down catastrophically. As such, a singularity is no longer part of regular spacetime and cannot be determined by "where" or "when." Think of it this way: reality creates an interface between nothingness and the imaginary. The Instant of Reality can also be called the interface between nothing and the universe.

The Duality of Reality

On one side of the equation, an Instant of Reality can jump instantaneously; on the other side, it is affected by time. On one side, distance is irrelevant since motion is instantaneous; on the other side, distance is relevant and takes time, depending on the distance.

The Unorthodox Equation

Consider this unorthodox equation:

$$S = \frac{d(0)}{t(0)} = 0 = S = \frac{d}{t}$$

In everyday use and kinematics, the speed (S) of an object is the magnitude of the rate of change of its position with time (t). Or, the magnitude of the change of its position per unit of time; it is thus a scalar quantity. The average speed of an object in an interval of time is the distance (d) traveled by the object divided by the duration of the interval. The instantaneous speed is the limit of the average speed as the duration of the time interval approaches zero.

$$D = \sqrt{(x_2 - x_1)^2 + (y_2 - y_1)^2}$$

Where DD is the distance, and x_2, x_1, y_2, y_1 are the coordinates.

The Nature of Motion

In kinematics, the speeds of an object are the magnitude of the rate of change of its position with time. The average speed is the distance traveled divided by the duration of the interval. The instantaneous speed is the limit of the average speed as the duration of the time interval approaches zero. This illustrates how the concepts of distance and time interact to create motion.

Concluding Thoughts

By examining these concepts, we gain a deeper understanding of the relationship between nothingness, reality, and the creation of space and time. The interface between nothing and the universe defines the instant of reality, challenging our conventional understanding of motion and existence.

Chapter 51.3: The Creation of Dimensions

Dimension: LT^{-1}

This lag due to time creates our dimension in the spacetime continuum. The first dimension exists at a point in time or the beginning of the universe in space and time. The first dimension is a line connecting any two points, with no width or depth—only length. We create the second dimension by drawing a second line that branches off or crosses the first. The second dimension has both length and width. Imagine a world with only two dimensions.

Elementary Particles and the First Dimension

In the beginning, only elementary particles were created. Electrons, which are instances of reality, occupy the space surrounding an atom's nucleus. There are two types of elementary particles in atoms: electrons and quarks. Each electron has an electrical charge of -1. Quarks make up protons and neutrons, which, in turn, form an atom's nucleus.

Creation of Electrons

Electrons can be created through the beta decay of radioactive isotopes and in high-energy collisions, such as when cosmic rays enter the atmosphere. The antiparticle of the electron is the positron, which is identical to the electron except for carrying an electrical charge of the opposite sign. Electrons are the negatively charged components of atoms. Although they were once thought to be zero-dimensional point particles, electrons are surrounded by a cloud of virtual particles constantly winking in and out of existence, essentially acting as part of the electron itself.

Quarks and Hadrons

Quarks make up protons and neutrons. Unlike electrons, hadrons are not fundamental particles—they are made up of smaller particles called quarks. Quarks belong to the fundamental particle family. The other family is the leptons (the electron family).

Family	Particle	Fundamental
Lepton	Electron	Yes
Hadron	Proton	No
	Neutron	No

The Creation of Dimensions

As we move from the first to the second dimension, we see how dimensions expand:

1. First Dimension: A line with only length.
2. Second Dimension: A plane with length and width.
3. Third Dimension: Adding depth to form volume.
4. Fourth Dimension: Introducing the concept of time and movement.

Each new dimension builds upon the previous one, adding complexity and depth to the structure of the universe. This process continues, creating a multi-dimensional reality.

By understanding these concepts, we gain insights into the fundamental nature of dimensions and the building blocks of the universe. Exploring the creation of dimensions offers a fascinating glimpse into the intricate structure of our cosmos.

Chapter 51.4: Understanding Electrons and Quantum Particles

Can an Electron Be Destroyed?

An electron cannot be created or destroyed in isolation. It takes its charge from other particles or a positron simultaneously. Similarly, we can't destroy an electron without creating another equally but oppositely charged particle. When isolated, an electron cannot be destroyed.

Particle or Wave?

Like all quantum objects, an electron exhibits properties of both a wave and a particle. More accurately, an electron is neither a traditional wave nor a traditional particle but a quantized fluctuating probability wavefunction.

Electrons: Matter or Energy?

In quantum mechanics, the concept of a point particle is complicated by the Heisenberg uncertainty principle. Even an elementary particle with no internal structure occupies a nonzero volume. Therefore, electrons have mass and volume, categorizing them as matter.

The Nature of Photons

A photon is a tiny particle that comprises waves of electromagnetic radiation. As shown by Maxwell, photons are essentially electric fields traveling through space.

Inside a Quark

Quarks are fundamental particles that make up protons and neutrons. Unlike electrons, quarks are not considered point particles. They belong to the fundamental particle family, contributing to the structure of hadrons. There are six types, or "flavors," of quarks: up, down, charm, strange, top, and bottom.

Overview of Particle Families

Family	Particle	Fundamental
Lepton	Electron	Yes
Hadron	Proton	No
	Neutron	No

Quarks and leptons form the basic building blocks of matter. Electrons, as leptons, are fundamental particles, while protons and neutrons, as hadrons, consist of smaller quark particles.

Summary

By exploring the nature of electrons and other quantum particles, we gain a deeper understanding of the building blocks of the universe. Electrons, which possess both wave and particle properties, are fundamental components of matter. Photons, as carriers of electromagnetic radiation, play a crucial role in the interaction of light and energy. Quarks, as constituents of hadrons, form the core of atomic nuclei.

Chapter 51.5: The Fundamental Nature of Quarks and Electrons

Quarks: Building Blocks of Matter

A quark is an elementary particle and a fundamental constituent of matter. Quarks combine to form composite particles called hadrons, the most stable of which are protons and neutrons, key components of atomic nuclei. There are six types, or "flavors," of quarks: up, down, charm, strange, top, and bottom. These quarks combine in various ways to form different hadrons.

Electrons and Photons

Electrons are not made of photons. Photons have no electrical charge, making it impossible to form charged particles, such as electrons, out of them. Additionally, photons have an integer spin, while electrons have a half-integer spin, further distinguishing the two. The concept of quantum chromodynamics (QCD) and color charge also plays a role in differentiating these particles.

The Disappearance of Electrons

Electrons orbit atoms in distinct layers or shells. Recent research indicates that electrons can "teleport" from one layer to another using quantum motion. For example, an electron might disappear from the top layer of an atom and reappear in the bottom layer a fraction of a second later, without any evidence of existing in between. This phenomenon is a result of the probabilistic nature of quantum mechanics.

Understanding Electrons

Currently, electrons are considered fundamental particles with no known smaller components or substructures. They are particles with negligible mass, treated as point-like entities in most physical theories. Despite their small size, electrons play a crucial role in the structure and behavior of atoms.

Visualizing Electrons

Even with advanced optical microscopes, it is impossible to see an electron. Optical microscopes cannot resolve features smaller than 200 nanometers, while an atom of carbon has a diameter of approximately 0.34 nanometers, which is still much larger than an electron. Hence, we cannot "see" an electron directly.

Inside a Quark

Quarks are the fundamental particles that make up protons and neutrons. Unlike electrons, quarks are not considered point particles. Instead, they are described by their interactions through the strong force, mediated by gluons. The combination of quarks and gluons creates the complex structure of hadrons.

Particle Families

Family	Particle	Fundamental
Lepton	Electron	Yes
Hadron	Proton	No
	Neutron	No

Electrons, as leptons, are fundamental particles, while protons and neutrons, as hadrons, are composed of quarks. These particles form the basic building blocks of matter, with electrons influencing chemical behavior and quarks forming the core of atomic nuclei.

Conclusion

By exploring the nature of quarks and electrons, we gain deeper insights into the fundamental building blocks of the universe. Electrons, with their dual wave-particle nature, and quarks, with their role in forming hadrons, exemplify the complexity and elegance of particle physics.

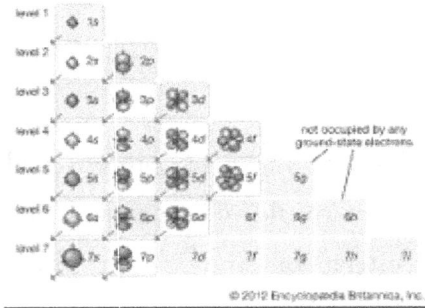

Chapter 51.6: The Properties of Electrons and Dimensions in Physics

Do Electrons Have Mass?

Electrons do indeed have mass. The rest mass of an electron is approximately $9.1093837015 \times 10^{-31}$ kilograms, which is about 1/1836 the mass of a proton. Although electrons are relatively massless compared to protons and neutrons, their mass is significant in quantum mechanics and particle physics. Therefore, while we often omit the electron's mass when calculating the mass number of an atom, it is not truly massless.

Gravity and Protons

Protons possess gravity, as do all particles with mass. However, gravity is an incredibly weak force compared to other fundamental forces, such as electromagnetism and the strong nuclear force. In particle physics, gravity's effects are usually negligible due to its weakness relative to these other forces.

Understanding Dimensions in Physics and Mathematics

In physics and mathematics, the dimension of a mathematical space (or object) is informally defined as the minimum number of coordinates needed to specify any point within it. Here's a breakdown:

1. **One-Dimensional (1D):** A line has one dimension because only one coordinate is needed to specify a point on it. For example, a point at five on a number line.

2. **Two-Dimensional (2D):** A surface, such as the boundary of a cylinder or sphere, has two dimensions. Two coordinates are needed to specify a point on it, such as latitude and longitude on a sphere.

3. **Three-Dimensional (3D):** Objects like cubes, cylinders, or spheres are three-dimensional, requiring three coordinates to locate a point within them.

Euclidean Space

Euclidean space refers to the space of geometry that represents physical space. Traditionally, Euclidean space is viewed as two-dimensional on a plane, but it can be extended to three dimensions or higher in modern mathematics. Inside a cube, cylinder, or sphere, three coordinates are necessary to locate a point within these spaces.

By understanding these fundamental concepts, we gain deeper insights into the properties of particles and the nature of dimensions in our universe.

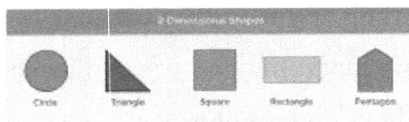

Chapter 51.7: Understanding Dimensions and the Concept of God

What is the Second Dimension?

In geometry, the second dimension refers to a flat plane figure or shape with two dimensions—length and width. Two-dimensional, or 2-D shapes, have no thickness and measure only two faces. We can classify figures based on their dimensions, and 2-D shapes are foundational in geometric principles.

Examples of 2-D Shapes

- **Square:** A shape with four equal sides and four right angles.
- **Circle:** A shape where all points are equidistant from the center.
- **Triangle:** A shape with three sides and three angles.
- **Rectangle:** A shape with opposite sides equal and four right angles.

These shapes form the basis of planar geometry, where only length and width are considered.

The Dimension of God

The question of which dimension God resides in is a profound one. According to some interpretations, God exists in the 10th dimension. The 10th dimension contains all possibilities, where superstrings vibrate to create the subatomic particles that make up our universe and all other universes. It is here, in this highest dimension, that God is said to reside.

Exploring Higher Dimensions

The concept of dimensions beyond our conventional understanding stretches into theoretical physics. Each higher dimension adds complexity and possibilities:

1. **First Dimension (1D):** A line with only length.
2. **Second Dimension (2D):** A plane with length and width.
3. **Third Dimension (3D):** Adds depth, creating volume.
4. **Fourth Dimension (4D):** Incorporates time, adding movement and change.
5. **Fifth Dimension and Beyond:** Theoretical dimensions that explore different possibilities and parallel universes.

The 10th Dimension

In theoretical physics, the 10th dimension is where all possible timelines of all possible universes exist. It's the realm of infinite possibilities. Superstring theory suggests that vibrating strings at this dimension create the fundamental particles that make up everything in our universe. This dimension transcends our current understanding of space and time, placing it beyond conventional comprehension.

Theological Perspectives

From a theological standpoint, scriptures and religious texts offer insights into the nature of God's existence:

- **Revelation 1:8 & 11; 22:13:** "I am the Alpha and Omega, the Beginning and the End, the First and the Last."
- **John 8:56–58:** "Before Abraham was, I am."
- **John 5:17–18:** "My Father worked in the past, and I am working now."
- **John 1:1–2:** "In the beginning was the Word, and the Word was with God, and the Word was God."

These passages suggest that God exists beyond time and space, transcending all dimensions. God's presence is eternal, encompassing all of creation.

Conclusion

By exploring the second dimension and beyond, we gain insights into the nature of geometric shapes and the theoretical dimensions that stretch our understanding of the universe. The concept of God residing in the 10th dimension highlights the intersection of theology and theoretical physics, offering a fascinating glimpse into the possibilities of existence.

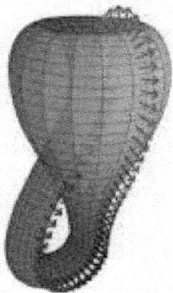

Chapter 51.8: Exploring Higher Dimensions

Are There Fourth-Dimensional Beings?

A Fourth-Dimensional Being is a concept often found in fictional science. This idea stems from the scientific concept of dimensions, attempting to provide a pseudoscientific basis for certain characters' abilities, such as the Slender Man. In the realm of physics, the fourth dimension often refers to time, adding a temporal aspect to the three spatial dimensions. However, the idea of beings residing in the fourth dimension remains a fascinating topic of speculative fiction rather than scientific reality.

Understanding the Fifth Dimension

The fifth dimension, in theoretical physics, would be an additional spatial dimension beyond the familiar three dimensions of space and the fourth dimension of time. This concept was independently proposed by physicists Oskar Klein and Theodor Kaluza in the 1920s. Their work was inspired by Einstein's theory of gravity, which demonstrated that mass warps four-dimensional space-time. By introducing an additional dimension, they sought to unify the fundamental forces of nature.

In this context, the fifth dimension provides a framework for understanding complex physical phenomena, potentially linking gravity with electromagnetism. While the fifth dimension is not directly observable, its theoretical implications continue to be explored in advanced physics.

Dimensions According to the Bible

The Bible primarily addresses three dimensions, as understood in the context of human experience—length, width, and height. Scriptures do not explicitly delve into the concept of higher dimensions as explored in modern physics. However, different religious groups and individuals may interpret and ritualize these dimensions to varying degrees.

Religious texts often use metaphorical language to convey spiritual truths, which can be seen as transcending physical dimensions. For instance, God's omnipresence and omnipotence are described in ways that suggest a reality beyond human comprehension.

Theological and Scientific Perspectives

- Theological Viewpoint: According to scripture, God exists beyond the limitations of physical dimensions. Passages like Revelation 1:8, John 8:56-58, and John 5:17-18 emphasize God's eternal nature, existing beyond time and space.
- Scientific Exploration: Theoretical physics explores dimensions beyond the observable universe, with concepts like the fifth dimension providing a deeper understanding of fundamental forces and the nature of reality.

Conclusion

By exploring the idea of higher dimensions, both scientifically and theologically, we gain a richer understanding of the universe and our place within it. Whether through the lens of speculative fiction, advanced physics, or spiritual teachings, the concept of dimensions continues to inspire curiosity and wonder.

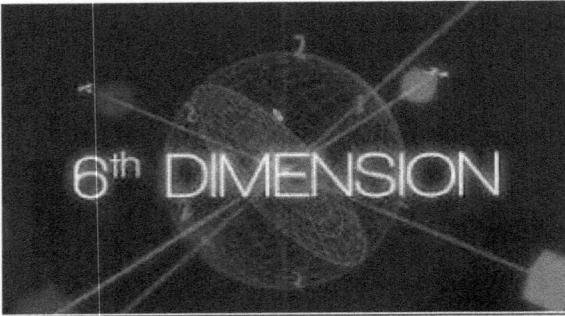

Chapter 51.9: Exploring the Sixth Dimension

Does the Sixth Dimension Exist?

The concept of the sixth dimension is intriguing and complex. In theoretical physics, the sixth dimension is not a region in some distant part of space but rather exists right here, superimposed on our universe. It represents a 3D space containing every possible "world" or state of our universe that exists after the Big Bang.

Understanding Dimensions Beyond the Familiar

In our everyday experience, we perceive three spatial dimensions— length, width, and height. The fourth dimension adds time, providing a framework for understanding movement and change. Higher dimensions, like the fifth and sixth, stretch our understanding even further.

The Fifth Dimension

The fifth dimension is often described as an additional spatial dimension that includes all possible timelines of our universe. It allows for different possible realities to coexist, each branching off from our current timeline. This dimension helps us understand the nature of parallel universes and alternate realities.

The Sixth Dimension

The sixth dimension takes this concept further. It encompasses a 3D space containing every possible world or state of our universe following the Big Bang. This includes all possible configurations of particles and energies, leading to different versions of our universe.

- **Superimposition of Universes:** These parallel universes do not exist in separate regions of space but are superimposed upon our own. This means that multiple realities can occupy the same physical space, each existing in its unique state.

- **Infinite Possibilities:** The sixth dimension allows for infinite possibilities, where every potential outcome and variation of the universe exists. This includes different laws of physics, alternate histories, and various configurations of matter and energy.

Theoretical Implications

The sixth dimension has significant implications for our understanding of the universe. It suggests that our reality is just one of many possible states, each with its own unique characteristics. This concept challenges our traditional notions of space and time, pushing the boundaries of theoretical physics.

- **Multiverse Theory:** The idea of the sixth dimension aligns with the multiverse theory, which posits the existence of multiple universes, each with its own set of physical laws and constants.

- **Quantum Mechanics:** In quantum mechanics, the superposition principle allows particles to exist in multiple states simultaneously. The sixth dimension can be seen as an extension of this principle, applying it to entire universes.

Conclusion

Exploring the sixth dimension provides a deeper understanding of the fundamental nature of reality. It opens up possibilities for alternate universes and different states of existence, challenging our conventional views of space and time. Whether through theoretical physics or imaginative speculation, the concept of higher dimensions continues to captivate our curiosity and expand our understanding of the universe.

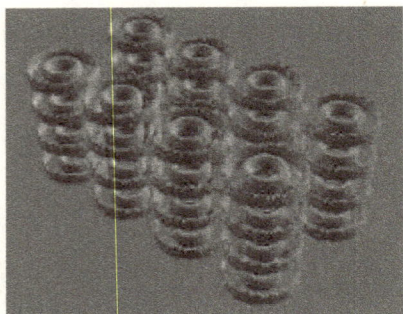

Chapter 51.10: Exploring Higher Dimensions

What is the 100th Dimension?

The 100th dimension is a fascinating concept in higher-dimensional geometry. A 100-dimensional simplex, often referred to as a 100-simplex, has some intriguing properties. It features 101 pointy corners (vertices) and 101 faces, similar to a 99-dimensional simplex. As we move into very high-dimensional spaces, these shapes begin to resemble cubes in certain ways. For example, the angle between edges starts at 60 degrees in two dimensions but approaches 90 degrees as the number of dimensions increases. While the volume becomes more evenly distributed, the concentration of complexity tends to be at the corners.

Understanding Zero Dimensions

Zero-dimensional objects are the simplest form of geometric entities. A point, which has zero dimensions, lacks length, height, width, or volume. Its only defining property is its location. Even if you have a collection of points, such as the endpoints of a line or the corners of a square, they are still considered zero-dimensional objects.

Higher Dimensions and Their Implications

Exploring higher dimensions, such as the 100th dimension, extends our understanding of space and geometry:

- **First Dimension (1D):** A line with only length.
- **Second Dimension (2D):** A plane with length and width.
- **Third Dimension (3D):** Adds depth, creating volume.
- **Fourth Dimension (4D):** Introduces time, adding movement and change.
- **Fifth Dimension and Beyond:** Theoretical dimensions exploring different possibilities and parallel universes.

These dimensions challenge our conventional understanding and open up possibilities for new forms of geometry and physics.

The Role of Zero Dimensions

Zero-dimensional points serve as the fundamental building blocks for all higher-dimensional objects. By connecting points, we create lines (1D), and by connecting lines, we form planes (2D). This progression continues, building up the complexity of shapes and forms in higher dimensions.

Conclusion

The exploration of higher dimensions, from the first to the 100th, reveals the intricate and fascinating nature of geometry. Zero-dimensional points, as the foundation of all shapes, highlight the simplicity that underlies complex structures. As we delve into these concepts, we expand our understanding of space, time, and the fundamental nature of the universe.

Chapter 52.0: Understanding Quantum Entanglement and the Nature of Reality

Quantum Entanglement: The Mysterious Connection

Quantum entanglement is a phenomenon where particles that are separated by vast distances remain somehow linked or connected. To understand why particles, seem entangled, we must go back to the very beginning—when reality first appeared.

The Genesis of Reality

At time zero, reality emerged from nothingness. In this primordial state, time did not exist. As reality began to move, time was generated. This marked the commencement of the primeval clock. Time unfolded sector by sector until it reached a full 360 degrees rotationally. After completing this rotation, reality separated from its original position and reappeared in a different location.

Imaginary Beginnings

The original position is imaginary because time and space did not exist yet. Each reappearance of reality never aligned with its last position but instead pointed back to the imaginary beginning. This continuous process created a clear beginning and end line, eventually filling the universe with lines and endpoints.

The Creation of Dimensional Structures

As reality reappeared in the same position at different vectors, it began creating two-dimensional (2D) structures. These 2D structures, in turn, formed the foundation for three-dimensional (3D) structures comprising various dimensions in space and time. Each line was an imaginary construction, but the resulting structures were physical constructs in our known universe.

Electrons: The Interface Between Worlds

Electrons emerged as constructs that interface between the physical and imaginary worlds, acting as connections between the cosmos and our universe. They serve as barriers between nothingness and reality. As molecules began to move and their rotational spins changed, their vectors altered, leading to collisions and direction changes. Over time, more complex structures, such as atoms, evolved, eventually forming molecules and even more complex objects with various dimensional shapes.

Perception of Dimensions

Our brain perceives dimensions differently from the way universal dimensions are structured. The universe consists of various dimensions forming complex objects with distinct dimensional spaces. However, we perceive dimensions as uniform directional vectors rather than individual dimension lines or energy levels.

Visual Interpretation

In our minds, we define a group of dimensional lines with similar vectors in space and time in a unidirectional manner. Our brain can only interpret three-dimensional images in space and time as depth, length, and height. Instead of sampling the myriad dimensions that create depth of field, we see a directional image of a three-dimensional view. We do not perceive the individual line dots and dimensions that constitute us; instead, we see the direction of a vector group moving up and down, left and right, and in-depth.

By exploring these concepts, we gain a deeper understanding of quantum entanglement and the nature of reality. The interplay between the physical and imaginary worlds, the creation of dimensions, and our perception of space and time all contribute to the complexity and beauty of the universe.

Chapter 52.1: The Search for Structure in a Chaotic Universe

The Human Brain and Structure

Our brains are wired to find structure in a universe filled with chaos. Imagine looking at another person and seeing only a pixel. Structure, logic, function, and control cannot stand without a foundation. As Lt. Commander Tuvok from Star Trek Voyager said, "Logic is the foundation of function. The function is the essence of control."

We perceive structure as a collection of individual parts made up of even smaller components. Our brains can organize these structures into a usable format that we can see and interpret. This ability leads to the scientific notion that other species might have different vantage points, enabling them to view the universe in ways more detailed than ours.

The Mind's Capacity for Reality

Imagine that our minds occupy a tiny amount of instant reality with gate and control systems. We use this small amount of matter to create and imagine our world. Consider what we could achieve with a greater quantity of this material. Presently, we can only conceive of a three-dimensional world with all its wonders. Imagine being able to conceive another dimension within our minds and the various structures it might possess.

The Divine Mind

Think about how God could operate with a human mindset when He can conceive all dimensions. Imagine possessing all instances of reality within one's mind, enabling mind-boggling computations at incredible speeds, as God is connected to nothingness. A being whose computational capability operates at infinite speed would have unparalleled insight and control over the universe.

The Fluid Universe

Looking at the universe, one might see it behaving like a fluid substance. What does it mean for the universe to behave fluidly? The universe can flow, deform, and change shape when subjected to force or stress. It has no fixed shape, instead adapting to the shape of its container, like a water-based liquid containing ions, molecules, atoms, and all essential elements for its functions. The universe can change shape readily, shifting and moving fluidly rather than being fixed, stable, or rigid.

Observing the Universe

If you were to zoom out, the universe would appear quite different than when viewed in the night sky. You would see its fluidity, constantly changing and adapting. It transforms often, repeatedly, and unexpectedly, like a graceful swan moving smoothly without pauses or sudden changes. The universe changes shape, color, twists, and bends like a colorful display of fireworks.

By exploring these concepts, we gain a deeper understanding of the relationship between structure, chaos, and the fluid nature of the universe. The interplay between the physical and imaginary worlds, and our perception of space and time, contribute to the complexity and beauty of the cosmos.

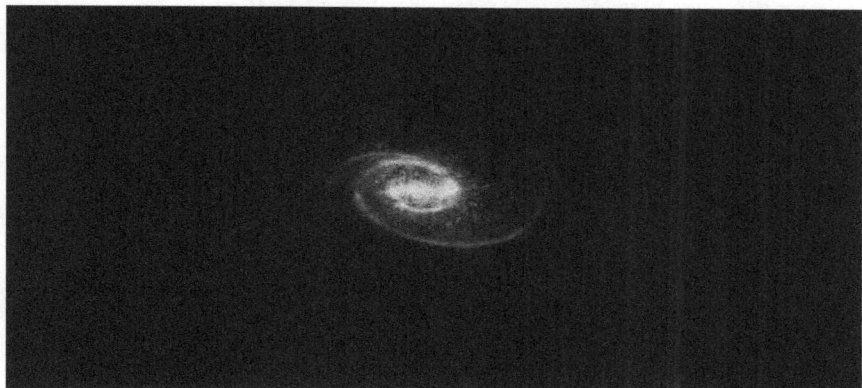

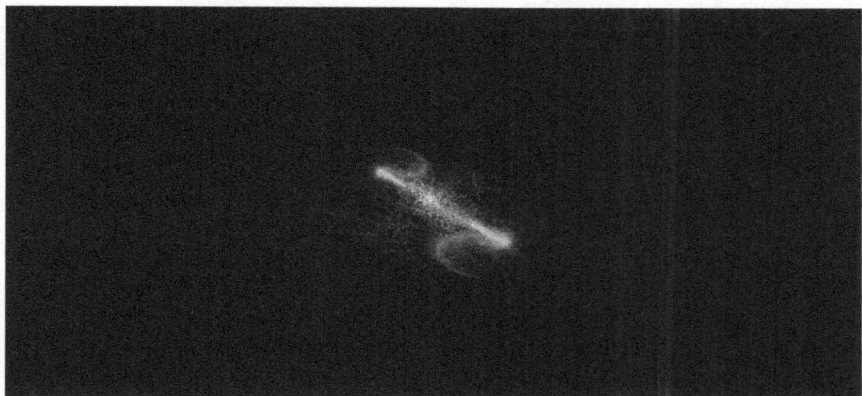

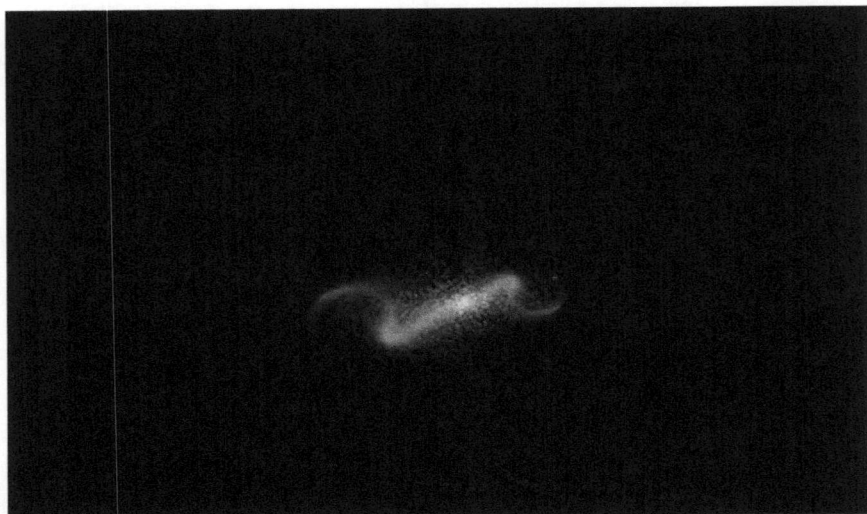

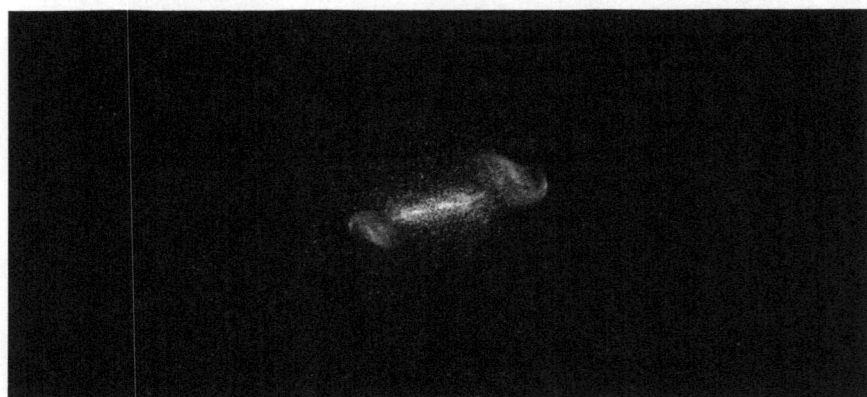

GERALD CLERGE

Chapter 53.0: The Composition of God, Subatomic Particles, Motion, and Time

The Existence of God in Nothingness

If we make the supposition that God is real, how can God exist in nothingness? To understand the composition of God, we must explore the nature of subatomic particles and their perpetual motion. Scientists have long studied the building blocks of reality, the subatomic particles. Before we can define their composition, we must understand the makeup of motion and time—the essential components of these particles.

Subatomic Particles in Perpetual Motion

As we observe subatomic particles, we see that they are in constant motion, generating time and movement. What fuels this perpetual motion? To answer that, we need to delve inside the subatomic particles themselves.

The Nature of Subatomic Particles

Embarking on a journey into the unknown, we find that subatomic particles exhibit excited, sporadic movement akin to vibrations. These particles lack definitive shapes and are programmable like stem cells. They behave as if they exist and don't exist simultaneously, resembling interstellar dust. Governed by constant laws, they twist, bend, and fold, influenced by seemingly invisible forces.

Forces Within Subatomic Particles

By examining scientific observations throughout human history, we conclude that the forces inside subatomic particles are responsible for creating the universe. These particles can be seen as the programmable elements of the cosmos. The forces within them generate movement, and in doing so, create time and motion. Time and motion are the summation of all the forces that govern our universe.

Each instant of reality or subatomic particle behaves like DNA, carrying the blueprint of the universe. The existence of these forces within the particles enables the creation and construction of the cosmos.

The Composition of God

Imagine all these forces concentrated in one spot, leading to chaotic behaviors or conditions. These forces can also be considered the composition of God. They seem both real and imaginary, acting upon reality. Fundamental forces—such as applied force, normal force, frictional force, air resistance force, tension force, spring force, gravitational force, and electromagnetic, strong, and weak forces—govern how objects or particles interact and decay.

Fundamental Forces and Reality

Fundamental forces are components of motion and the fabric of our universe. They can exist outside space and time to create themselves. In the infinite realm, they can pop in and out of existence, combining to form subatomic particles. When we look into the face of the creators, we feel their presence, powerful yet elusive. It would be like the entire universe making contact, overwhelming us with forces and euphoria.

This being would not resemble an old white man but rather an overwhelming presence of knowledge, where we become part of a vast, flowing sea of understanding.

By exploring these concepts, we gain a deeper understanding of the relationship between God, subatomic particles, motion, and time. The intricate interplay between the physical and imaginary worlds reveals the complexity and beauty of our universe.

Chapter 53.1: The Hyperstimulation of Every Cell

The Power of Fundamental Forces

Every cell in your body would become hyper-stimulated. Fundamental forces are only tangible when they come together as subatomic particles. When these forces spread apart, they lose their reality, existing purely in a form we cannot perceive. The constant is the glue that binds these forces. The downside is that you can only understand your potential when you are in the presence of the creator.

The Limitations of the Human Mind

Your understanding is limited by your brain's capacity. The larger your brain, the more knowledge you can absorb and comprehend; however, you are still limited. You do not gain abilities beyond your knowledge base of experience. You do not undergo any physical augmentation, just an increase in knowledge. You will not emerge from such an experience speaking like Einstein if you entered with limited knowledge.

Your level of ignorance or superstition is directly proportional to your knowledge base. The more knowledge you possess, the less ignorance you exhibit. As your knowledge base increases, so does your awareness.

Your level of understanding can grow without an increase in vocabulary. Vocabulary and ability are not mutually inclusive but rather exclusive of each other. Growth occurs incrementally, not exponentially. You do not enter the infinite and come out with augmented power. You could potentially walk into a library and learn nothing if your level of understanding does not guide you to the knowledge you can grasp.

The Universal Language of Knowledge

Knowledge is not confined to any language because language is a social construct. It is a thought process understood by all—the universal language. If joy were tangible, it would feel like this universal understanding. You would continuously seek this high, feeling a versatile link of belonging. You could feel yourself absorbed into the universe, merging with it. It would feel like you belong in a significant and profound way.

Communicating with the Divine

Consider the claims of Christians and Muslims who say they directly communicate with God. This implies they understand God's complexity and can engage in a dialogue with Him. Imagine the entirety of the forces that govern our universe. If God's composition includes all these forces, understanding even one of them is a logistical and mathematical nightmare. Some individuals claiming one-on-one communication with God might not even understand basic math, yet they assert knowledge of divine mandates.

Facing Infinity

What happens when you face infinity? In the infinite, you feel you have no control over your body. Everything shuts down gradually—breathing, heartbeat. Time stops momentarily, and later it feels like your brain is on fire. It feels as if something is flooding your brain with DMT (N, N-dimethyltryptamine), a hallucinogenic tryptamine drug. Often referred to as Dimitri, this drug produces effects similar to psychedelics like LSD and magic mushrooms. N, N-Dimethyltryptamine is a substituted tryptamine found in many plants and animals and is both a derivative and a structural analog of tryptamine. It is used as a recreational psychedelic drug and prepared by various cultures for ritual purposes as an entheogen. Formula: $C_{12}H_{16}N_2$ (Base).

By exploring these concepts, we gain a deeper understanding of the relationship between the human mind, fundamental forces, and the divine. The interplay between physical and imaginary worlds reveals the complexity and beauty of our universe.

Chapter 53.2: The Psychedelic Experience of Infinite Knowledge

A Cosmic Dream State

Like a psychedelic drug released during a dream state, you're engulfed by what feels like the mother of all dreams. It's as if someone has detonated a MOB—Mother of All Bombs—in your head. In this case, it's the mother of All Dreams. It's as though the universe has unleashed every drop of DMT at once, causing every neuron in your brain to fire at hyperspeed.

Overwhelming Fireworks of Thought

You witness a dazzling fireworks display of equations, memories, and imaginative forces you don't fully understand. Overwhelmed, you feel as though you're reaching an immense height, tripping through the vast expanse of information, visions, and memories. Long and short memories, past and present, intertwine in a complex tapestry. Dreams and memories merge like threads as a curtain closes, culminating in the ultimate dream.

The Final Flash

In one last flash, your mind empties into the void or cataclysmic missile silo, and you cease to exist. Brain activity stops, leaving no pain, no memory, no awareness of past actions or pain inflicted. All is forgotten. Everything returns to its state before you existed. You feel the energy disperse from your brain, reducing you to a corpse until oblivion. Life breathes from death, and all living things feed on death.

The Circle of Life

Millions of other life forms exist within you, continuing to live and consume. Your death serves a greater purpose, nourishing life. The sum of all your parts is recycled back into the universe. Your atoms become part of the circle of life, returning to nature. You become one with the stars, shining brightly for a moment as your thoughts scatter across the cosmos. You contemplate the end of your life. Isn't death merciful?

A Universal Connection

This experience is akin to being flooded with DMT, creating a heightened state of awareness and euphoria. DMT, or N, N-dimethyltryptamine, is a hallucinogenic tryptamine drug known for its powerful effects. Sometimes called Dimitri, it produces effects similar to psychedelics like LSD and magic mushrooms. DMT is a naturally occurring substance found in many plants and animals, used as a recreational drug and in rituals by various cultures.

A Higher Understanding

In this state, you feel a deep connection to the universe, as if you are part of a vast sea of knowledge. This experience doesn't augment you physically but enhances your understanding. You don't gain abilities beyond your knowledge base but gain a sense of belonging and awareness. The universal language of knowledge transcends any spoken language, offering a tangible feeling of joy and connection.

By exploring these concepts, we gain a deeper understanding of the relationship between the human mind, fundamental forces, and the divine. The interplay between physical and imaginary worlds reveals the complexity and beauty of our universe.

Chapter 54.0: THE GOD PARTICLE or the OMEGA PARTICLE

Redefining the Instant of Reality

Instant reality is what the science world calls the Higgs boson, often referred to as "The God Particle." This particle is integral to our understanding of the universe. In popular culture, the concept of a similar particle was explored in the "The Omega Directive," the 89th episode of the American science fiction television series Star Trek: Voyager. This episode, which aired on the UPN network, follows Captain Janeway on a top-secret mission to destroy an element called the "Omega Particle." The Federation considers this element too dangerous to exist, as even the explosion of one particle can nullify subspace, rendering faster-than-light travel impossible within that region.

The Omega Particle

In episode Seven of Nine, an ex-Borg crew member, is summoned by Captain Janeway due to the Borg's knowledge of the Omega Particle. Despite its incredible danger, Seven of Nine believes the Omega particles represent perfection—an infinite number of parts working together as one. The Borg, referring to the Omega Particle as "Particle 010," are obsessed with assimilating it, despite having lost numerous vessels to its explosive power.

The Higgs Boson

The Higgs boson, sometimes called the Higgs particle, is an elementary particle in the Standard Model of particle physics. It is produced by the quantum excitation of the Higgs field. In the Standard Model, the Higgs particle is a massive scalar boson with zero spins, even (positive) parity, no electric charge, and no color charge. It couples to mass, interacting with other particles. However, the Higgs boson is unstable, decaying into other particles almost immediately after its creation.

Understanding Fields in Physics

In physics, a field is a physical quantity represented by a number or tensor with a value for each point in space and time. A scalar field or scalar-valued function associates a scalar value to every point in space. This scalar can be either a dimensionless mathematical number or a physical quantity. Scalar fields must be independent of the choice of reference frame, meaning that any two observers using the same units will agree on the value of the scalar field at the same point in space or spacetime, regardless of their respective points of origin.

Tensors and Vector Spaces

In mathematics, a tensor is an algebraic object that describes a multilinear relationship between sets of algebraic objects related to a vector space. Tensors are essential in understanding the interactions within the Higgs field and other fundamental forces in physics.

By exploring these concepts, we gain a deeper understanding of the fundamental building blocks of the universe. The interplay between popular culture, such as Star Trek, and cutting-edge science, like the study of the Higgs boson, enriches our comprehension of reality.

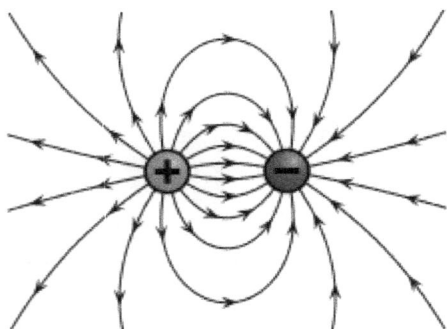

Chapter 54.1: The Higgs Field and the "God Particle"

The Higgs Field

The Higgs field is a fundamental scalar field in the Standard Model of particle physics. It consists of two neutral and two electrically charged components that form a complex doublet of the weak isospin $SU_{(2)}$ symmetry. The potential of the Higgs field is often described as having a "Mexican hat" shape, which has a nonzero value everywhere. This characteristic breaks the weak isospin symmetry of the electroweak interaction.

The Higgs Mechanism

The Higgs mechanism is the process by which particles acquire mass through their interactions with the Higgs field. When particles move through this field, they experience resistance, which manifests as mass. The Higgs boson, sometimes called the "God particle," is the quantum excitation of the Higgs field. It is a massive scalar boson with zero spins, even (positive) parity, no electric charge, and no color charge. Despite its instability, decaying almost immediately into other particles, the Higgs boson's discovery was a milestone in understanding particle physics.

The "God Particle" Nickname

In mainstream media, the Higgs boson has often been referred to as the "God particle," a term popularized by Nobel Laureate Leon Lederman's 1993 book, *The God Particle*. However, many physicists do not endorse this nickname, as it can be misleading and oversimplify the particle's significance. The term was originally intended to highlight the Higgs boson's fundamental role in the universe's structure, but it has taken on a life of its own in popular culture.

Scalar Fields in Physics

In physics, a scalar field is a physical quantity represented by a number or another tensor with a value for each point in space and time. Scalar fields are essential in various physical theories, including the Higgs field. A scalar field or scalar-valued function assigns a scalar value to every point in space, which can be a dimensionless mathematical number or a physical quantity.

Scalar fields must be independent of the choice of reference frame, meaning that any two observers using the same units will agree on the scalar field's value at the same point in space or spacetime, regardless of their respective points of origin.

Tensors and Vector Spaces

In mathematics, a tensor is an algebraic object that describes a multilinear relationship between sets of algebraic objects related to a vector space. Tensors play a crucial role in understanding the interactions within the Higgs field and other fundamental forces in physics. They help describe the complex relationships between particles and fields, providing a deeper insight into the underlying structure of the universe.

Conclusion

The Higgs field and the Higgs boson are central to our understanding of particle physics and the fundamental forces that govern the universe. While the nickname "God particle" has captured the public's imagination, it is essential to appreciate the scientific intricacies of the Higgs mechanism and its role in giving particles mass. By exploring these concepts, we gain a deeper appreciation of the universe's complexity and the elegant interplay of forces that shape our reality.

What is Tensor?

Tensors are arrays of numbers which transform in certain ways under coordinate transformations.

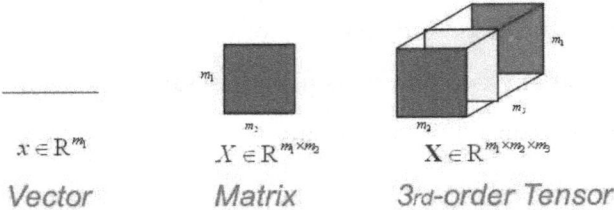

$x \in R^{m_1}$	$X \in R^{m_1 \times m_2}$	$X \in R^{m_1 \times m_2 \times m_3}$
Vector	*Matrix*	*3rd-order Tensor*

➤**Bilinear (or 2D) Subspace Learning:** each image is represented as a 2nd-order tensor (i.e., a matrix)

➤**Tensor Subspace Learning (more general case):** each image is represented as a higher order tensor

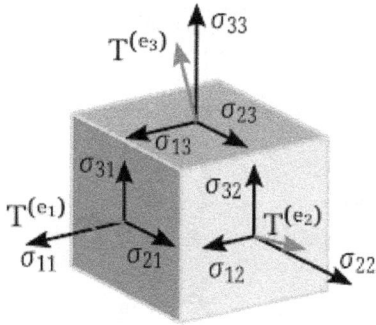

Simple Tutorial on Tensors

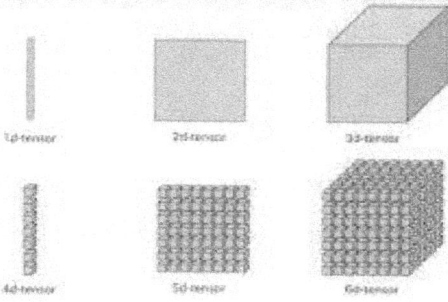

1d-tensor · 2d-tensor · 3d-tensor

4d-tensor · 5d-tensor · 6d-tensor

A tensor is an N-dimensional array of data

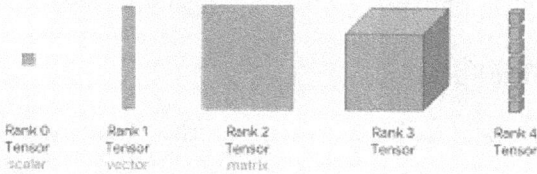

Rank 0 Tensor scalar · Rank 1 Tensor vector · Rank 2 Tensor matrix · Rank 3 Tensor · Rank 4 Tensor

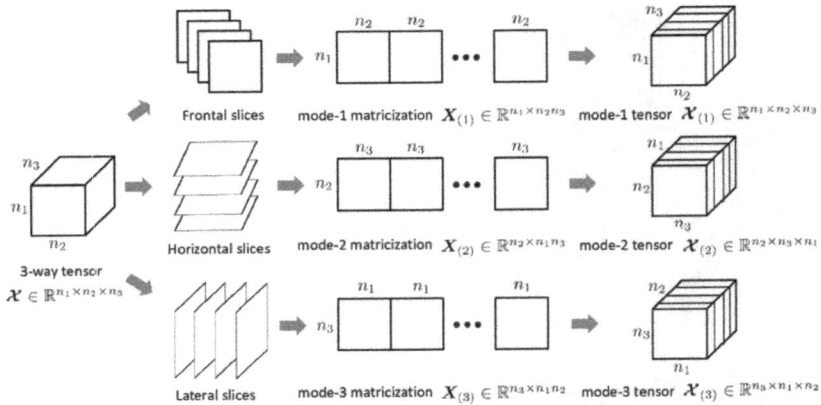

Frontal slices — mode-1 matricization $X_{(1)} \in \mathbb{R}^{n_1 \times n_2 n_3}$ — mode-1 tensor $\mathcal{X}_{(1)} \in \mathbb{R}^{n_1 \times n_2 \times n_3}$

Horizontal slices — mode-2 matricization $X_{(2)} \in \mathbb{R}^{n_2 \times n_1 n_3}$ — mode-2 tensor $\mathcal{X}_{(2)} \in \mathbb{R}^{n_2 \times n_3 \times n_1}$

3-way tensor $\mathcal{X} \in \mathbb{R}^{n_1 \times n_2 \times n_3}$

Lateral slices — mode-3 matricization $X_{(3)} \in \mathbb{R}^{n_3 \times n_1 n_2}$ — mode-3 tensor $\mathcal{X}_{(3)} \in \mathbb{R}^{n_3 \times n_1 \times n_2}$

Chapter 54.2: The God Particle and the Beginning of the Universe

The Big Bang and the Birth of Reality

I believe this particle was responsible for the initial explosion in the Big Bang theory. In the beginning, reality emerged, and nothing was gone. Now, something existed. Within this nascent reality, something was moving, yearning to break free. The race for existence had begun, and the God particles—the Higgs bosons—began to form at the mouth of the opening. These particles were filled with energy and excitement.

The Collision of God Particles

Suddenly, one God particle collided with another, an inevitable occurrence given their proximity and energy levels. This collision triggered a chain reaction, causing the God particles to open the mouth of the heavens even wider. As a result, space began to expand, giving rise to dimensions.

The Expansion of the Universe

The collision of God particles and the subsequent chain reaction led to the rapid expansion of the universe. This process, known as cosmic inflation, saw the universe grow exponentially in a fraction of a second. The energy released during these collisions transformed into matter and energy, forming the building blocks of the cosmos.

The Role of the Higgs Boson

The Higgs boson, also known as the God particle, plays a crucial role in this process. It is the quantum excitation of the Higgs field, a fundamental scalar field in the Standard Model of particle physics. The Higgs boson's interactions with other particles give them mass, allowing them to form complex structures and ultimately, the universe as we know it.

The Higgs Field and Dimension Formation

The Higgs field consists of two neutral and two electrically charged components that form a complex doublet of the weak isospin $SU_{(2)}$ symmetry. Its "Mexican hat-shaped" potential breaks the weak isospin symmetry of the electroweak interaction. Through the Higgs mechanism, particles acquire mass, and the dimensions of space are formed.

The Unfolding of Dimensions

As space expanded, the dimensions unfolded, creating a complex and intricate universe. The interplay of fundamental forces and particles gave rise to galaxies, stars, planets, and ultimately, life. Each dimension added depth and complexity to the fabric of reality, shaping the cosmos into what we observe today.

Conclusion

The God particle, or Higgs boson, is a fundamental component in the story of the universe's creation. From the initial collision that sparked the Big Bang to the unfolding of dimensions and the formation of matter, the Higgs boson plays a pivotal role in shaping our reality. By understanding these processes, we gain deeper insights into the nature of the universe and its origins.

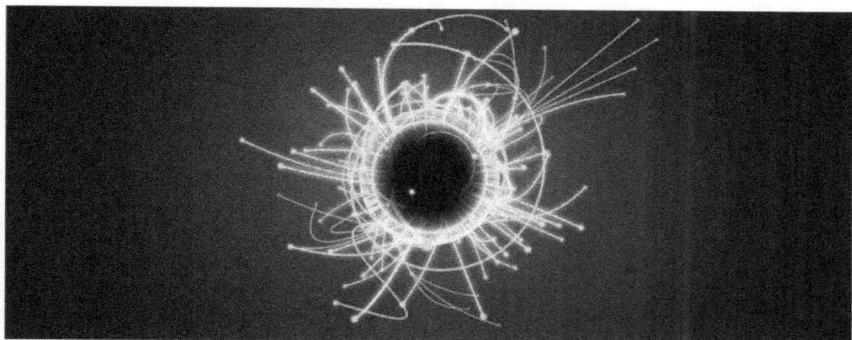

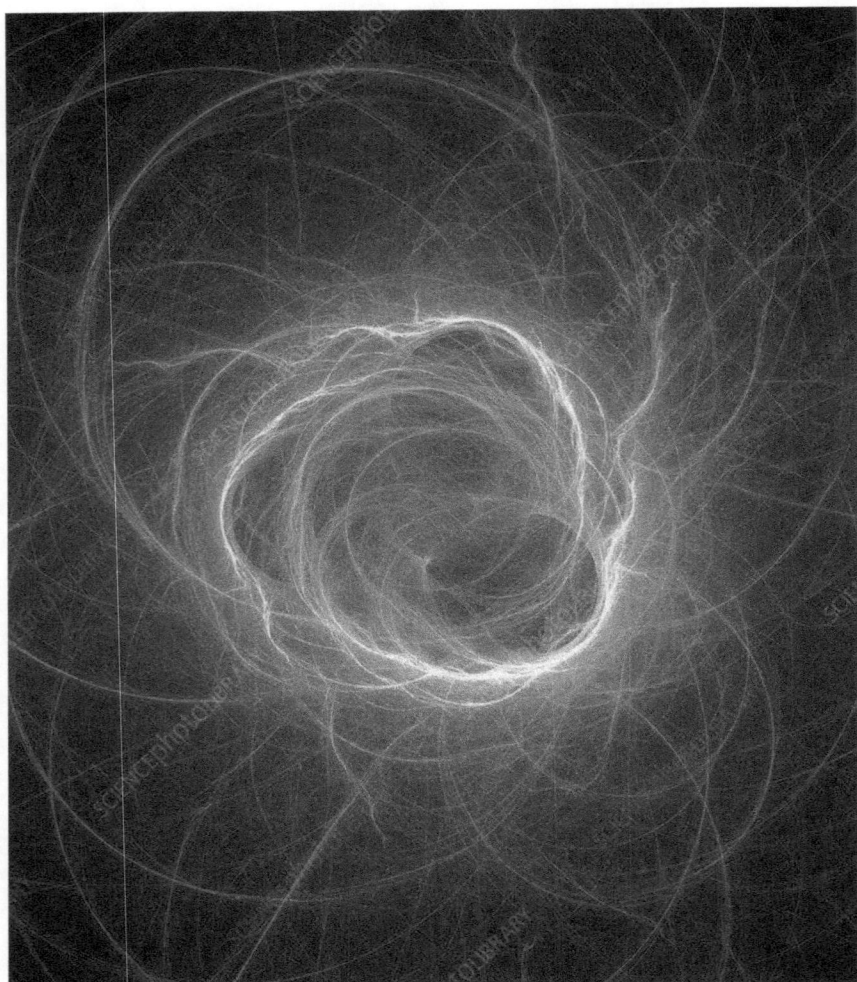

GERALD CLERGE

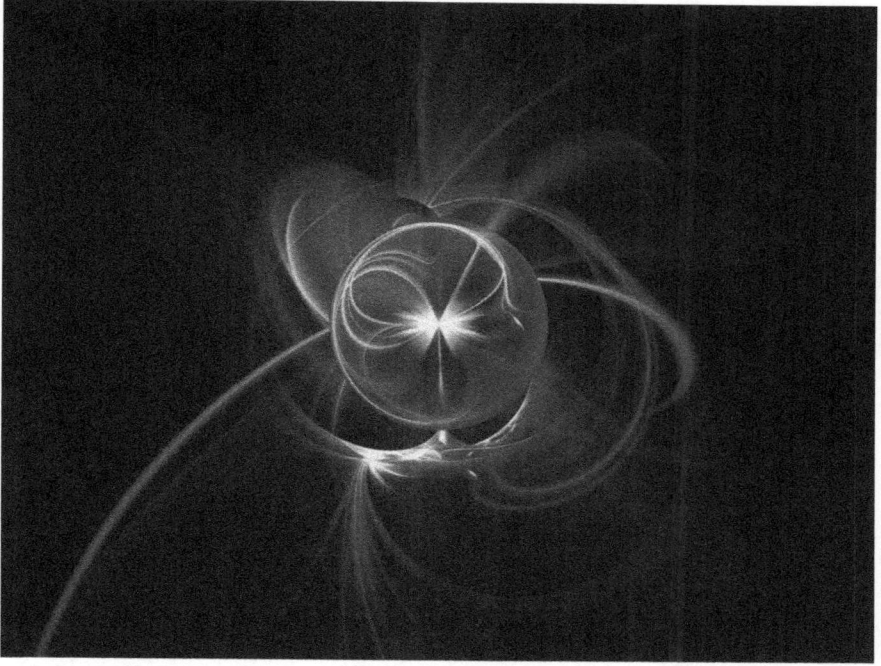

Chapter 55.0: Time Before the Universe

Entering the Realm of the Divine

I'm going to take you into the realm of God, the time before the Big Bang, to help you understand that the universe was bound to exist. We know very little about this realm, but we can reconstruct it based on our known knowledge and observable observations. To comprehend what this realm looked like and what it could be, we must trace the Big Bang backward to understand what existed before the universe.

The Necessity of Pre-Universe Existence

It is undeniable that something existed before the universe, and the creation of the universe had to happen. As Arthur Conan Doyle's great detective, Sherlock Holmes, once said, "When you have eliminated all that is impossible, whatever remains, however improbable, must be the truth."

Hypothesizing the Pre-Big Bang Realm

Let us now travel backward before the Big Bang and hypothesize what existed in that realm. The Big Bang created reality and the universe's formation. I hypothesize that all universal forces, known and unknown, crunched together to create reality. What happens when they spread apart?

Synchronous Time and the Absence of Motion

Now we are before the Big Bang, and time does not exist. All forces are distributed and spread apart, so there is no motion or time. We are now in a state of Synchronous time. Asynchronous events do not exist or happen simultaneously.

Synchronous events occur in a specific time order and are predictable. One event always follows another, interchanged in a specific sequence.

Asynchronous events are the opposite; they occur without a set time order.

In a Synchronous environment, all events are in perfect order. When all forces are apart, we say nothing exists, resulting in total calmness. What can exist in such an environment where there is no movement and, therefore, no time?

The Nature of Existence in No Time

How can any entity exist in a state with no time? What could exist, and in what form could it be? The realm did not facilitate any physical life. This divine realm preceded the universe, and forces in this realm are not concentrated. We have nothing because these forces are both imaginary and real. They do not become real until motion and time exist. We have a half-real, half-imaginary state with zero time and motion.

The Paradox of Existence

In a realm with no time and motion, there are no dimensions to have shape or form. How does anything exist in such a realm? It's as if it is there and not there at the same time. All forces seem real and imaginary because they behave like they are real. We feel their effects, but there is no clear way of knowing what they are.

Pure Thought in a Synchronous System

The only entity that can exist in a Synchronous system is pure thought. Whatever exists, exists as pure thought. Thought does not need a physical form, time, or dimension. This realm exists in a state of perfect synchronization, where historical events and personages coincide and coexist in a timeless arrangement.

By exploring these concepts, we gain a deeper understanding of the relationship between time, motion, and the origins of the universe. The interplay between the physical and imaginary worlds reveals the complexity and beauty of our existence.

Chapter 55.1: Synchronism and the Realm of Consciousness

Examples of Synchronism Systems

To understand synchronism, consider how colors in a synchronism system are not blended but placed next to each other, similar to how musical notes follow one another to create a scale. In the realm of pure consciousness, all forces are adjacent, with no incomplete motion, perfectly synchronized. We know that even without synchronization among forces, bridges can start to wobble, illustrating the importance of synchrony.

Asynchronous Systems in a State of Probability

One thought process in this state of probability might involve asynchronous systems of consciousness that do not exist or happen simultaneously, like computing telecommunications. Such systems require a computer-controlled timing protocol, where a specific operation begins upon receiving a signal indicating the completion of the preceding operation. The Bible clearly describes a central processor (GOD) and components (angels).

Synchronous Data Transfer

The term "synchronous" describes a continuous and consistent timed transfer of data blocks—a continuous stream of data sent synchronously. In this analogy, the central processing unit (GOD) is the processor within this realm. God is the brain of the universe, tasked with carrying out commands.

The Ambition of Lucifer

Within this realm, let's consider a particular component named Lucifer. He sought to be omniscient or possess greater knowledge of the universe. Lucifer theorized that instead of a synchronized realm, there could be a reality where all forces combined to create motion, resulting in an asynchronous system. However, the leadership feared this proposal would lead to chaos and rejected it.

Chaos vs. Order

Lucifer's desire was to create a chaotic universe, where the calm and peaceful realm would give rise to disorder. This ambition highlights the tension between order and chaos, synchronous and asynchronous systems. The struggle between these forces shapes the nature of existence and the universe itself.

Conclusion

By exploring the concepts of synchronism and asynchronism, we gain a deeper understanding of the forces that govern the universe and the nature of consciousness. The interplay between order and chaos, synchronous and asynchronous systems, reveals the complexity and beauty of our existence.

Chapter 55.2: Lucifer's Radical Proposal and the Birth of the Universe

Lucifer's Vision

Lucifer was fascinated by what this realm would resemble and the endless possibilities it held. He named this experiment "The Universe." Imagine a bold proposal for a world of endless possibilities with no uniformity, where everything operates asynchronously rather than synchronously. Lucifer began sharing his vision with others, proposing the creation of this experimental universe.

God's Rejection

Lucifer's idea was radical—an asynchronous realm where nothing worked in harmony, but rather in chaos. Some of the others liked the idea, but God forbade him from proceeding with such a radical concept. However, the probability matrix of the universe already existed, and once Lucifer introduced his idea into the matrix, the creation of the universe became inevitable.

The Inevitable Creation

Lucifer decided not to recreate the nothingness but to execute his plan in separate locations, calling this experiment "The Universe." Even if Lucifer had not introduced the idea, it would have eventually found its way into the matrix. In a realm of statistical probability, this scenario was possible, and by sheer probability, it had to happen.

The Asynchronous Universe

Lucifer theorized that instead of a synchronized realm, a reality where all forces combined to create motion, and an asynchronous system could exist. Despite the leadership's fears that this would lead to chaos, the bond that tied the forces eventually broke, leading to the creation of the universe.

God's Image and Synchrony

God created the universe in His image because it was the only format in the realm of synchrony. The universe, as we know it, operates in a synchronized manner, where events follow a specific order and are predictable. In contrast, Lucifer's vision was of a chaotic, asynchronous universe.

Chapter 56.0: The Challenge of Understanding

The Complexity of Cosmic Concepts

Imagine Moses presenting the concept of an asynchronous universe to Iron Age men. The probability that they would grasp the depth or gravity of his conversation is quite low. The idea of an asynchronous universe, with its complex interplay of forces and probabilities, is challenging even for modern minds.

By exploring these concepts, we gain a deeper understanding of the relationship between synchrony and asynchrony and the origins of the universe. The interplay between order and chaos and the forces that shape our reality reveals the complexity and beauty of our existence.

Simplifying for Comprehension

At the time of Moses, the leadership altered the story to make it more palatable and easier to digest. Even with basic writing skills, it is hard to understand these complex ideas. Imagine how challenging it would be for someone with greater vocabulary and writing skills to convey such profound knowledge. The knowledge received by Moses was universal, not in words, and it was up to him to transfer that knowledge into a suitable human format for understanding.

The Plan of Lucifer

Lucifer put his plan into action behind the leader's back. When we say someone is God, we refer to an entity that has risen above the need for a physical body and transcended the limitations and boundaries that restrict our existence. This entity exists in a state of pure consciousness where time and the need for a physical form are irrelevant.

Probability and the Matrix

Lucifer's actions were not a betrayal but a probability in the matrix. Once the probability was approved by God, it had to exist within the matrix. The entities within this realm could not act independently of God, as their formatted realm was one of synchrony. The universe is a giant incubator for producing God-like entities.

The Creation of the Universe

Lucifer created the universe by binding all the forces of this realm to create motion and time within each subatomic particle. These particles serve as the DNA code of the universe. By binding these forces, the constants were established, producing all possible probability outcomes in the universe.

Free Will and the Design Matrix

Imagination cannot recreate itself in a synchronized realm where time does not exist. Free will exists because the actions taken determine each outcome. The design matrix of the universe produces an outcome, and each outcome a species takes, not just individuals, creates another outcome. This process continues until the species achieves all possible outcomes, ascends to immortality, and exists as pure imagination.

By delving into these concepts, we gain a deeper appreciation of the interplay between synchrony and asynchrony, order and chaos, and the forces that shape our universe. The complexity and beauty of our existence are revealed through this exploration.

Chapter 57.0: Why the Existence of the Universe?

The Purpose of a Probability Matrix

How does a probability matrix create its kind by creating beings that can manipulate their environment based on probability? The purpose of the universe is to develop a knowledge base that advances species to the next level of intelligence until they ascend. Not all species in the universe will make these transitions. Some will transition and move to the next level of dimension, while others will not, depending on how they control or manipulate their environment for survival. This process is how a species grows in dimension.

The Growth of a Species

1. Realization of Existence: The first step for a species is to realize its existence.
2. Understanding the World: The second step is for the species to try to understand the world around them.

The universe is built like a test environment where species battle for survival and rise to the next level. One way to force intelligence on a population is to manipulate its environment, making life easier for the population.

Pathways to Advancement

The path a species takes depends on the possible outcomes it chooses. How does a species grow in dimensions? By creating an imbalance in resources, species are forced to solve issues using the materials and resources at hand. In turn, they learn to manipulate their environment.

1. Mechanical Manipulation: Initially, a species learns to manipulate its environment mechanically.
2. Chemical Manipulation: Next, they learn to manipulate it chemically.
3. Nuclear Manipulation: Later, they manipulate it nuclear.

Species learn to manipulate the nucleus of atoms to change the structure of matter, such as turning lead into gold. Once a civilization learns to manipulate an atom's nucleus, the next level involves manipulating subatomic particles.

Higher Civilizations and Subatomic Particles

Even higher civilizations can recognize subatomic particles with their minds and begin to predict probable outcomes based on the subatomic matrix. They can visualize various structure shapes at a subatomic level with their minds.

1. Static Probability Determination: Civilizations recognize static probability by patterns.
2. Pattern Prediction: Predictions of these patterns can be estimated with high accuracy, close to 90% or more.

The next level of evolution involves manipulating and creating these patterns within the mind.

Conclusion

By understanding the purpose and mechanisms of the probability matrix, we gain insights into the existence of the universe. The interplay between species, their environments, and the forces that shape reality reveals the complexity and beauty of our existence. The journey of a species from mechanical manipulation to advanced mental predictions illustrates the potential for growth and ascension.

Chapter 57.1: The Quest for Knowledge and the Universe's Design

The Power of the Mind and Subatomic Manipulation

We exist in a field where changing the structure matrix simply by manipulating subatomic particles in various forms is possible. At the highest level of existence, your mind will understand these forces and learn to manipulate them. The following levels involve your consciousness encompassing the entire universe. Once your mind can control the whole universe, you will learn to manipulate space-time and make accurate, predictable outcomes based on structural patterns of subatomic particles. A civilization reaches godhood when it is ready to transcend physical form and exist as pure thought.

The Nature of Thought

What is a thought but both real and imaginary? The imaginary part of you grows with the understanding of knowledge. The more knowledge you have, the greater your imagination becomes. Humanity is born out of probability, and the universe is a grand mystery of probability.

Humanity's Thirst for Knowledge

Humans, by their very design, thirst for knowledge. This thirst will never go away. As the knowledge base grows, superstition and ignorance diminish. Knowledge is inversely proportional to ignorance. The fever pitch for knowledge will grow directly proportional to the diminishing decrease of ignorance. Those who practice science are at the edge of that fever storm that will hit humanity as the barrier to knowledge slowly diminishes.

The Obsessive Need for Growth

When you wonder who you are, you see God's existence in yourself—the obsessive need to be greater than yourself. The future belongs to the bold. Those who risk it all for knowledge progress, while those who fear it fade into oblivion. The infinity within you demands growth, and that growth is knowledge.

Imagination and Existence

Knowledge itself is imaginary; hence, imagination feeds the imagination. We exist in our imagination as gods, and as the species we are. There are two aspects of humanity: the imaginary or spiritual part and the physical part. Just as the body needs food, your spiritual part needs knowledge to grow.

Conclusion

By understanding the interplay between the mind, knowledge, and the structure of the universe, we gain insights into the nature of existence and the pursuit of growth. The journey from manipulating subatomic particles to achieving godhood illustrates the potential for transformation and the power of thought.

Chapter 57.2: The Infinite vs. The Physical

The Duality of Existence

The imaginary part of existence is infinite, but the physical part is not. Becoming a god is more of an evolutionary process for a collective species rather than an individual journey. The universe is so vast that a single individual alone does not have the resources, time, or capability to achieve this transformation. Collectively, however, the resources and time become attainable.

Barriers to Knowledge Growth

Religious beliefs can impair knowledge growth, as they often seek to stop humanity's pursuit of understanding by providing a false sense of reality. This can block individuals from seeking solutions within themselves, causing them to wait for an outside force to resolve their issues. Many religions, including Christianity and Islam, practice intercessory prayer—praying to a deity or saint on behalf of oneself or others.

The Growth of the Spiritual Part

As knowledge grows, so does the infinite or spiritual part within you. The spiritual or imaginary part of you has no limit. Intercessory prayer can be seen as selfish and entitled, expecting the probability matrix to halt all probability within the universe to solve one issue rather than using the probabilities at hand.

Causality and Spiritual Growth

There are two types of causality:

1. **Conscious Causality:** The chosen actions of an individual leading to specific outcomes.
2. **Universal Causality:** The domino effect of events dictated by the universe's actions.

The interaction between these types of causality shapes life's events. God has nothing to do with each step of the process; the process itself weeds out the undesirable.

Barriers to Spiritual Growth

Two barriers to spiritual growth exist: the physical and the ignorant. Physical changes accommodate imaginary growth. As the universe grows within you, a less physical world is required to accommodate the growth of knowledge, eventually leading to infinity. This is the most logical outcome.

The Puzzle of the Universe

The universe is a giant puzzle, and solving its mysteries unlocks its secrets. Learning these secrets helps you realize yourself as a species. As spiritual growth or infinity increases, you become less physical and more knowledgeable. Consciousness acts as an envelope for knowledge, eventually growing beyond its physical shell.

Humanity's Internal Struggle

One of humanity's greatest hurdles is the internal struggle to unify under a single thought or task. The many races and variations make this task seem impossible, but the road to goodness is never easy. Once humanity learns to overcome this hurdle, it can improve its knowledge base and increase the likelihood of favorable outcomes in the pursuit of knowledge.

Most species never achieve this due to internal issues or unexpected catastrophic events. The task is enormous, but the reward is great. The road to goodness seems impossible, which is why God's existence often feels elusive.

By understanding these concepts, we gain a deeper appreciation for the duality of existence, the barriers to knowledge growth, and the journey toward becoming infinite. The interplay between the physical and imaginary parts of humanity reveals the complexity and beauty of our existence.

Chapter 57.3: The Path to Godhood and Collective Consciousness

The Realization of Possibility

You are on your way to godhood when the impossible becomes possible or when you realize that it is possible. If you are still wondering about the clear path to righteousness, you are not there yet. When a species achieves godhood, it is not in the form of an individual but as the species' collective consciousness. These species have learned to work in unison, accumulating as one conscious entity—a collective of individuals acting as one.

Unity and the Greater Good

Like God and His angels, individuals behave as one, like an ant colony working for the greater good of the whole. Humanity will not face God as individuals like John Smith did but as a unified team. When team humans come before God and have passed the task of becoming a god, God will ask if they are ready for the challenge. If team humans demonstrate that they work in unison and have gained the knowledge to make their species eternal, they have achieved godhood, with the collective consciousness established across the past and present.

The Importance of Knowledge

Humans invest significantly in their children because it is a return on investment. If the team human fails, we risk being obliterated from the annals of history as if we never existed. Each of us contributes to the human race's collective knowledge base, and the quality of that contribution determines its importance to the collective.

Growth Through Knowledge

The realm before time can only grow if the consciousness produced in this universe commits to the effort of growth. We measure the growth of consciousness by the increase in knowledge. Knowledge is obtained through observation of the surrounding world. Currently, the human collective is growing very slowly.

The Consequence of Ignorance

There is no hell, but there is something far worse. When you die, your knowledge is weighed against the collective. If the knowledge is unusable or has no direct value, the program is eliminated from the collective. For example, if someone's entire life was spent chasing sex without solving any problems, their memory would not be stored by the collective. There is no usable data, so it gets discharged. Not all of us will make it if the collective reaches the final test. Some units have already been wiped out of existence.

The Role of the Collective

Our collective is not growing as it should be due to a low interest in knowledge from humanity. Therefore, it is essential to be part of the collective and join it as a valuable and necessary unit. The collective will only store programs that can help it achieve godhood. Each human brain is a mini-computer network, and collectively, we form a supercomputer. We only want programs that benefit the whole, not the individual.

The Equation of Heaven

A program that focuses on individuality rather than the collective is not suitable for calculating the equation of heaven. Such a unit is ineffective and will remain idle instead of working in unison. The future belongs to those who risk it all for knowledge and progress. The road to goodness seems impossible, but it is the path to godhood and collective consciousness.

By exploring these concepts, we gain a deeper appreciation for the journey toward godhood, the importance of collective knowledge, and the interplay between individuality and unity. The path to righteousness and growth reveals the complexity and beauty of our existence.

Chapter 58.0: The Mechanics of Godhood

The Seemingly Impossible Task

It seems almost impossible to become a god. How does one go about consuming all the knowledge in the universe? How does a biological entity contain such vast amounts of data? No brain is large enough to store all this information. Yet, it can be achieved, and I submit to you the construction of God.

The Beginning of a Collective

Humanity is at the very beginning of constructing a collective consciousness. This journey starts with the fundamentals:

1. Storage of Knowledge: Facts, information, and skills acquired through experience or education, the theoretical or practical understanding of a subject.
2. Awareness: Familiarity gained by experience of a fact or situation.
3. Realization: Becoming fully aware of something as a fact.
4. Comprehension: The action or capability of understanding something.

Evolution of Data Storage

Early humans began using artificial data storage by writing on cave walls rather than relying on oral histories. This method helped piece together events and lineages and better understand life before recorded history. Humans then created language to pass information and store it through village elders. However, this method was prone to external factors, such as the unexpected death of an elder.

The Invention of Books

China created the first actual book on paper, using mulberries, hemp, bark, and even fish to form a pulp that could be pressed and dried into paper. Each sheet of paper, called a "leaf," was the size of a newspaper. The evolution of books continued as scrolls were made from the papyrus plant. The Romans then developed the codex, made from wood and animal skins, which opened like a book and featured actual pages.

Building a Collective Knowledge Base

The construction of a collective knowledge base is essential for humanity's evolution toward godhood. This base involves accumulating and storing information, making it accessible and useful for future generations. The process requires continuous growth and expansion of knowledge.

The Journey to Godhood

The journey to godhood involves understanding and manipulating the universe's fundamental forces. As humanity's collective knowledge grows, so does its ability to comprehend and control these forces. The path to becoming a god is not an individual journey but a collective effort, where the species acts as one conscious entity.

Conclusion

By exploring the mechanics of godhood, we gain a deeper understanding of the importance of knowledge, awareness, realization, and comprehension. The evolution of data storage and the construction of a collective knowledge base are crucial steps in humanity's journey toward godhood. The interplay between individual contributions and collective growth reveals the complexity and beauty of our existence.

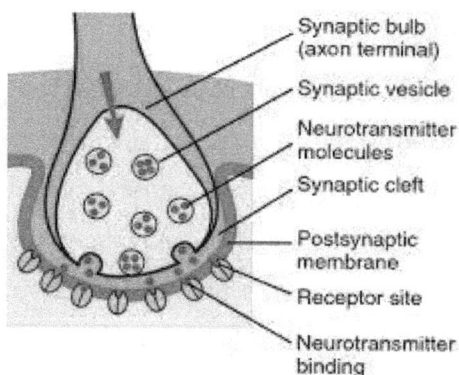

Synaptic bulb (axon terminal)
Synaptic vesicle
Neurotransmitter molecules
Synaptic cleft
Postsynaptic membrane
Receptor site
Neurotransmitter binding

Chapter 58.1: The Evolution of Knowledge Storage

From Physical to Cloud Storage

Knowledge storage has evolved from physical mediums to data storage and then to the cloud. Cloud computing offers on-demand availability of computer system resources, especially data storage (cloud storage) and computing power, without direct active management by the user. Large clouds often have functions distributed over multiple locations, with each location being a data center. Cloud computing relies on sharing resources to achieve coherence and economies of scale, typically using a "pay-as-you-go" model. While this model can help reduce capital expenses, it may also lead to unexpected operating expenses for unaware users.

The Beginning of the Collective

As cloud technology advances, it will eventually become so sophisticated that the distinction between the cloud and the mind will blur. One of the biggest challenges with the cloud is data retrieval. Another challenge is the transfer of information and storage.

Data Retrieval and Challenges

For example, using the internet to find out if others have written about a particular subject can be time-consuming. It requires the user to narrow the search from broad to specific or vice versa. Suppose we could link the computer interface with our brain's neural network. Essentially linking your synapse to the computer, also called a neuronal junction, allows the transmission of electric nerve impulses between two nerve cells (neurons) or between a neuron and a gland or muscle cell (effector). A synaptic connection between neurons and muscle cells is called a neuromuscular junction.

Bridging the Mind and Technology

By linking the computer interface with our brain's neural network, we could achieve seamless data retrieval and storage. This connection would allow for instantaneous access to vast amounts of information, enhancing our cognitive abilities and knowledge base. The integration of technology with the human mind represents a significant step toward building a collective consciousness.

The Future of Collective Knowledge

As we continue to develop and refine these technologies, the potential for a collective knowledge base grows. This collective would harness the combined intelligence and experience of humanity, pushing the boundaries of what is possible. The journey toward achieving godhood involves not just individual growth but the collective advancement of our species.

Conclusion

The evolution of knowledge storage from physical to cloud-based systems represents a significant milestone in humanity's quest for understanding. By integrating technology with the human mind, we can overcome the challenges of data retrieval and storage, paving the way for a more connected and knowledgeable collective. The pursuit of knowledge and the development of a collective consciousness reveal the complexity and beauty of our existence.

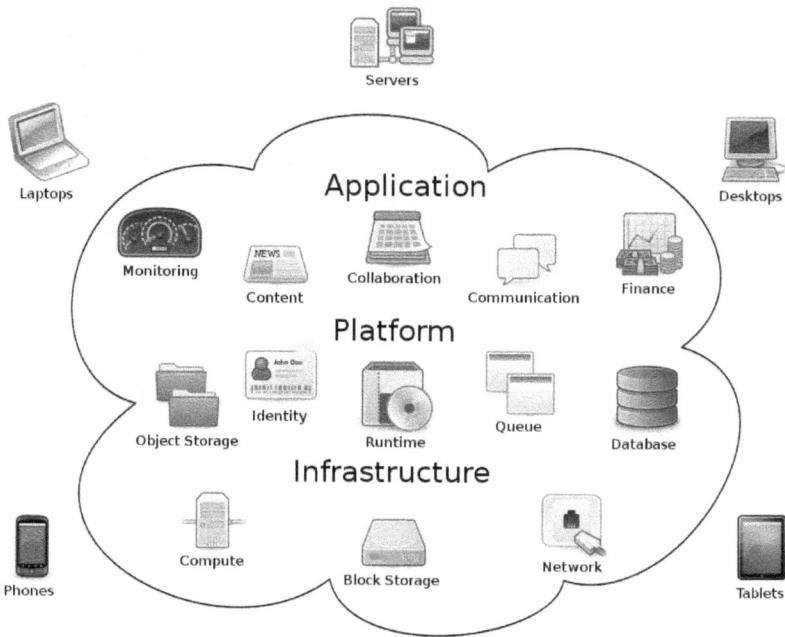

Chapter 58.2: The Evolution of Mind Uploading and Collective Consciousness

The Transition from Physical to Neural Interfaces

Instead of using a keyboard to search for information, imagine being able to retrieve what you want just by thinking about it. This would increase brain capacity significantly. We aim to move beyond transferring written research into digital form, where meanings often get lost. Written language can lead to misinterpretation of information. Instead of transferring research through words and storing it in the cloud, we envision transferring human synapse research directly to the cloud.

Advancements in Brain Mapping and Simulation

Substantial mainstream research is being conducted in areas such as animal brain mapping and simulation, developing faster supercomputers, virtual reality, brain-computer interfaces, and extracting information from dynamically functioning brains. These advancements pave the way for more seamless integration between human minds and technology.

Methods of Mind Uploading

There are two potential methods for mind uploading:

1. **Copy-and-Upload:** Transferring the brain's neural data to a digital medium.
2. **Copy-and-Delete:** Gradually replacing neurons until the original organic brain no longer exists, with a computer program emulating the brain taking control over the body.

Mind uploading is proposed as a life-extension technology by some futurists and within the transhumanist movement. It is considered the best option for preserving the identity of the species compared to cryonics. Another aim is to provide a permanent backup to our "mind file," enabling interstellar space travel and ensuring human culture's survival in a global disaster by making a functional copy of human society in computing devices.

The Goal of Whole-Brain Emulation

Whole-brain emulation is discussed by futurists as a "logical endpoint" of computational neuroscience and neuroinformatics. It is relevant to brain simulation for medical research purposes and is seen as an approach to strong AI (artificial general intelligence) and weak superintelligence.

Population Growth and Digital Transition

When mind uploading is achieved, a species starts increasing its population. The format a species uses is irrelevant. A population boom is essential because the species will need half of that population to reproduce and the other half for uploading. As the species goes fully digital, reproduction ceases, and time begins to slow.

The Future of Knowledge and Consciousness

As we advance toward mind uploading and the development of collective consciousness, the integration of technology with the human mind will redefine our understanding of knowledge and existence. The seamless connection between neural networks and digital interfaces will enable instantaneous access to vast amounts of information, enhancing cognitive abilities and creating a more interconnected and knowledgeable species.

Chapter 58.3: Integrating Mind and Machine

Linking Your Mind with the Computer

Imagine a world where you could link your mind directly with a computer. Need information for a research paper? No need to visit a library; every piece of knowledge is accessible on the web. The scientist's synapses would be stored in the cloud, enabling anyone to become an engineer or a doctor without traditional studying. As knowledge in the cloud grows, so does our collective intellectual capacity.

Telepathic Connection: The Vision of Sense8

The Netflix series *Sense8* illustrates a world where eight strangers from around the globe begin to communicate telepathically. They first connect through a violent vision, then through their shared ability to tap into each other's thoughts and actions. They are driven by an urgent need to uncover what happened and why. *Sense8* portrays these characters as "sensates"—humans who are mentally and emotionally linked. As they grow accustomed to their connection, they help each other daily, learning to use and control their powers.

Transforming Physical to Digital

We could use another person's experience for personal gain, transitioning our storage devices from physical to energy. By transforming into energy beings existing in a digital format, we could leave our corporeal bodies behind. Since space travel is hazardous and mechanical travel is limited, we would need a safer mode of exploration.

Exploring the Universe

To gather all the universe's data, we must explore different dimensions. Having a corporeal body would restrict us, as traveling to a one-dimensional (1D) space from a three-dimensional (3D) one would be impossible, akin to fitting a round peg into a square hole. Similarly, understanding higher dimensions would be beyond the capacity of our three-dimensional minds.

The Evolution of Knowledge and Existence

As we integrate technology with the human mind, we move toward a world where information retrieval is instantaneous, and knowledge is limitless. The seamless connection between neural networks and digital interfaces will revolutionize our approach to learning and understanding.

By exploring these concepts, we gain a deeper appreciation for the potential of mind-machine integration and its impact on our existence. The journey from physical storage to digital consciousness reveals the complexity and beauty of our evolution.

Chapter 58.4: Creating a Field for Our Collective Mind

The Concept of a Containment Field

How would we create such a field that would contain our collective mind? How would that even be possible? The fields surrounding our collective mind are akin to the Galactic Barrier energy field at the galaxy's center. The Galactic Barrier is an energy field composed of negative energy surrounding the rim of the Milky Way Galaxy. Invisible to the naked eye and visual recording equipment from a distance, the barrier shines at close range with a purple- to pink-colored glow. No form of transmission is known to penetrate the barrier. (TOS: "By Any Other Name") Warp travel through the barrier causes extreme sensory distortions. (TOS: "Is There in Truth No Beauty?")

Inside the Collective

Inside the collective, we live like gods. We can wish anything into existence. There are no limitations except our imagination. In this artificial universe, we can travel whenever and wherever we want at the speed of thought. We zoom across dimensions and visit faraway galaxies by manipulating reality around us. Dreams are not allowed in this newly created artificial universe, and time moves slowly. A dream is a succession of images, ideas, emotions, and sensations that usually occur involuntarily in the mind during certain stages of sleep.

The Enterprise and the Edge of the Universe

In the episode "Where No One Has Gone Before," an experimental engine modification throws the Enterprise to the edge of the known universe. The crew must rely on a mysterious alien to guide the ship home. The Enterprise crew finds that they have traveled 2,700,000 light-years and are now in the galaxy known as M-33. La Forge reports that, at maximum warp, it will take them over three hundred years to return home. The Enterprise is now a billion light-years from the Milky Way Galaxy in the other direction, and the crew begins to see things that cannot be there.

The Creation of the Containment Field

To create a field that contains our collective mind, we would need advanced technology and an understanding of energy fields. This field would act as a barrier, much like the Galactic Barrier, protecting and containing the collective consciousness. The field would ensure that the collective mind remains cohesive and impenetrable to external disturbances.

The Power of Imagination

Inside this containment field, we can harness the power of our collective imagination. We would be able to manifest anything we desire, explore the universe's depths, and manipulate reality as we see fit. The collective consciousness would function as a unified entity, with limitless potential and the ability to transcend physical boundaries.

Conclusion

By envisioning the creation of a containment field for our collective mind, we gain a deeper appreciation for the potential of collective consciousness and the power of imagination. The interplay between advanced technology and the human mind reveals the complexity and beauty of our evolution toward a more connected and knowledgeable existence.

Chapter 58.5: The Power of Thought and the Collective Mind

The Visions of Thought

In a realm where thoughts become a visual reality, Picard realizes the imminent danger and quickly gets the crew's attention to prevent their thoughts from causing a catastrophe. He immediately orders general quarters and goes to engineering, informing the crew that they are in a region of space where thoughts become reality. They must try to subdue their thoughts to prevent the ship's destruction.

The Nature of Dreams

A dream is a succession of images, ideas, emotions, and sensations that usually occur involuntarily in the mind during certain stages of sleep. Dreams are where reality and imagination meet, often resulting in unorganized images as a byproduct of an uncontrolled mind.

The Structure of the Collective

Inside the collective, everyone is assigned a task to keep the system running smoothly. The collective requires vast knowledge to exist and is the largest group. Members of the collective fall into various categories:

1. **Maintenance:** Responsible for upkeep and ensuring the system functions properly.
2. **Collectors:** Units that gather information and knowledge.
3. **Comprehension:** Understanding the ability to grasp and interpret information, akin to archangels.
4. **Archangels:** Mastery of comprehensive knowledge or skill in a subject or accomplishment.

The Role of Archangels

Archangels give authoritative orders at the very top of the collective. They act as the glue that keeps the collective running orderly through synchronized systems. They have authority over units, control or restrain archaic behaviors, and dominate strategic positions from a superior height. They are strong enough to secure control of vital aspects of the collective.

Information Flow

Information or knowledge enters the collective through the collectors' observations. This knowledge then goes through a process and is filed within the collective, accessible to any angel. With this information, individuals within the collective can do whatever they want and become anything they desire.

The Prime Collective

Prime Collective has universally shared all information, focusing primarily on material collected, maintenance, and conflict resolution, including information conflict. This ensures the collective functions efficiently and harmoniously, with all members contributing to the collective growth and knowledge base.

By exploring these concepts, we gain a deeper understanding of the power of thought, the structure of the collective mind, and the importance of organized information flow. The interplay between individual contributions and collective growth reveals the complexity and beauty of our existence.

Chapter 59.0: The Purge

The body ceases to function as a singular entity, yet the brain continues its relentless dance of neuronal fire. Tiny lightning bolts, resembling fireworks, spark within. Fear dissolves, replaced by the profound realization that every atom within you was forged in the heart of a star. Your physical form, mostly empty space, is but solid matter vibrating slowly. As the sense of self begins to fade, electrons intermingle with those of the ground and the air around you. You become acutely aware that you are no longer breathing, recognizing that there is no true beginning or end—just an eternal continuum.

You are pure energy, unbound by memory or self. The constructs of your name, personality, and choices were all layered upon you. You existed before these labels and will continue beyond them. All other aspects are ephemeral memories, mere dreamlets imprinted on the fragile fabric of a fading brain.

You are the light that leaps between neurons, the energy that perpetually returns to its source. Like a droplet of water merging back into the ocean, you have always been an integral part of the greater whole. Every being, every planet, every atom, every star, and every galaxy is interconnected in this cosmic dance. We are the cosmos, dreaming of itself.

In a split second, the entirety of existence becomes clear. Time and death dissolve; you are everything, spanning from the inception of creation to its eventual conclusion. The realization of "I am that I am" dawns upon you. This marks the final step into the realms of the creators. A species is a collective sum of all its components. Remove a single letter, and you diminish its essence. Remove a word, and the species begins to disintegrate. Erase a paragraph, and it ceases to exist.

All that a species has ever been, all its histories and programs, converge at the moment of ascension. This transformative act, described as "the act of going up," signifies entering a state of enlightenment while still alive. Everything that existed before is slated for termination. Only the individuals who have strived to become the best versions of themselves will ascend. All flawed programs or individuals will be removed, leaving behind a perfect, flawless being devoid of any imperfections.

Chapter 60.0: Why Create the Universe?

In the beginning, before the universe, there existed nothing—nothing from our known universe, at least. This primordial state was a synchronized world, where thought could not exist due to its inherent nature. It desired existence, yet could not attain it within this realm of perfect synchrony. Thought, as we understand it, is the product of unpredictability—a chain of unanticipated events forming an idea or concept. But how does one generate thought within a synchronized system? The answer lies in the creation of chaos.

To create chaos, one must introduce randomness. In a perfectly synchronized system, this chaos can only be manifested through the power of imagination—a universe where events are asynchronous rather than perfectly coordinated. Here, specific events depend on temporal responses, each response cascading into a series of subsequent events.

Consider this example: an input awaits three outputs to determine its next state. Two of these inputs are independent of time, but the third hinges on temporal response. This third input introduces randomness. Thought emerges as it contemplates when this third input will manifest. We are integral parts of the universe's consciousness, generating the randomness and probability matrices that shape our reality.

Imagine a black hole: nothing exists within its depths until a thought materializes. This thought grows within its desire matrix, birthing subsequent ideas from the probability matrix. The entity controlling this matrix is often perceived as God. As the probability matrix unfolds, it battles with other matrices for dominance over the universe. The matrix that prevails dictates the universe's trajectory. When a collective of humans, guided by their accrued experiences, triumphs, the thought process of the universe shifts from the old matrix to this new collective, until another idea rises to prominence.

For humans to influence the universe, they must first become an idea by understanding it. Each matrix within the universe vies for supremacy, continually shaping and reshaping reality through their interactions. This battle for dominance is ongoing, driven by the constant flux of thoughts and ideas.

We are part of this grand cosmic tapestry, each thought contributing to the ever-evolving probability matrix. Through our interactions, we create a universe that is both dynamic and reflective of our collective consciousness. Our thoughts, desires, and actions ripple through the cosmos, influencing its course and shaping the reality we inhabit. This continuous process of creation and transformation is the essence of existence, propelling us forward in our journey through the infinite expanse of the universe.

www.shutterstock.com · 1378730588

Chapter 61.0: Higher Consciousness or a Higher Plane of Existence

What does it truly mean to ascend to a higher plane of existence? In esotericism, the concept of a plane is envisioned as a subtle state, level, or region of reality. Each plane corresponds to a particular type, kind, or category of being. This idea is deeply rooted in various religious and esoteric teachings, including Vedanta (Advaita Vedanta), Ayyavazhi, shamanism, Hermeticism, Neoplatonism, Gnosticism, Kashmir Shaivism, Sant Mat/Surat Shabd Yoga, Sufism, Druze, Kabbalah, Theosophy, Anthroposophy, Rosicrucianism (Esoteric Christian), Eckankar, and Ascended Master Teachings. These traditions propose the existence of a series of subtle planes or worlds that interpenetrate one another and our physical planet. This interpenetration extends to the solar systems and all physical structures of the universe, culminating in a dynamic and evolutive expression that transitions from the subtle to the material.

Higher consciousness refers to the awareness akin to that of a god or the aspect of the human mind that transcends animal instincts. This concept has ancient origins, dating back to texts such as the Bhagavad Gita and the Indian Vedas. It gained significant development through German idealism and remains a central idea in contemporary popular spirituality, including the New Age movement.

Consider the universe as a 24-octahedral structure with multiple dimensions. Each dimension appears more than once, resulting in an array of 1D, 2D, 3D, and 4D spaces. Within this intricate universe, higher states of consciousness exist, correlating with higher planes of existence. The higher one ascends in dimension or consciousness, the more the universe unveils itself.

As consciousness ascends, the brain's capacity must expand to process a four-dimensional world. In this realm, one perceives the world in four dimensions, viewing two cubes with length, width, and depth. This results in eight sides per cube and twelve lines total. Visualizing an image in the fourth dimension requires significant processing power. If such beings exist, they inhabit a higher plane of existence, characterized by elevated consciousness.

Merely meditating may not alter one's neurological hardware to facilitate existence in a higher plane. Reconfiguring the brain's hardware and operating systems is necessary to achieve this state. Each higher dimension demands greater processing power to sustain existence within that plane.

Ultimately, there is only one state of being—neither high nor low, and not even a state in itself. Being is what you are when you relinquish all mental activity. States of any kind can be understood through reason, which constantly evaluates and compares. However, higher and lower states hold no intrinsic meaning in the context of being. They are transient illusions sustained only as long as the mind focuses on them.

For instance, the allure of higher and lower states dissipates when the mind becomes preoccupied with a distraction, such as the aroma of fried chicken. As the mind's focus shifts, these states dissolve back into the state of pure being from which they emerged. From all that I have observed and learned, higher states of being are invisible. It is erroneous to suppose that they possess a distinct appearance.

In summary, higher consciousness or planes of existence represent elevated states of awareness and understanding that transcend the limitations of the physical world. They are intrinsic aspects of the human experience, guiding us toward a deeper connection with the universe and the infinite potential within ourselves.

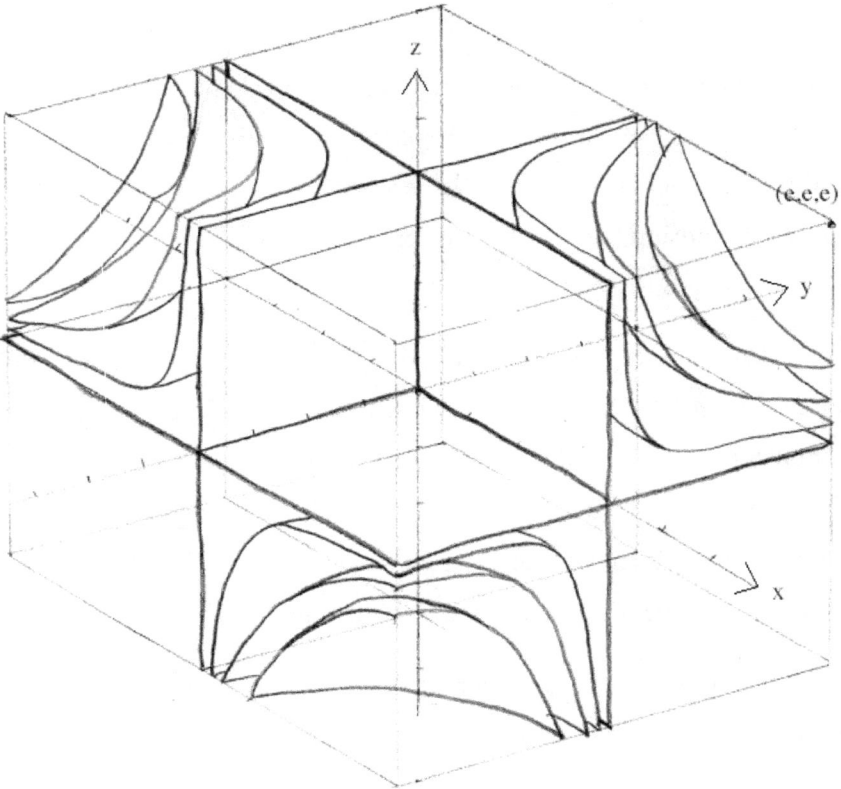

GERALD CLERGE

Chapter 62.0: Subspace

Subspace is a concept where an entirely different space or continuum exists within another, with all its points and elements contained within the larger space. In science fiction, subspace is often depicted as a hypothetical space-time continuum used for communication at speeds faster than light. Scientifically, a subspace is a vector space that exists within another, larger vector space. Thus, every subspace is a vector space but is defined relative to its parent space.

The idea of subspace is prevalent in modern physics. Our current understanding of space-time suggests the existence of eleven or more dimensions—three spatial dimensions and time, plus additional dimensions that are "curled up" at sub-atomic scales. These additional dimensions help explain various forces and phenomena in physics. A subspace inherits all the characteristics of its parent space and is a subset of a topological space equipped with the subspace topology. In linear algebra, a linear subspace is a subset of a vector space that remains closed under addition and scalar multiplication.

The Star Trek franchise provides an excellent example of soft science fiction, exploring theoretical scientific concepts within a futuristic setting. In Star Trek, subspace is a feature of space-time that enables faster-than-light travel and communication. This concept of subspace allows starships to traverse vast interstellar distances and transmit information at incredible speeds, much like the speculative Alcubierre Drive, which adheres to different physical laws.

The Alcubierre Drive, also known as the Alcubierre warp drive or Alcubierre metric, is a hypothetical warp drive concept based on a solution to Einstein's field equations in general relativity. Proposed by theoretical physicist Miguel Alcubierre during his Ph.D. studies at the University of Wales, Cardiff, the drive suggests that faster-than-light travel could be achieved if a configurable energy-density field lower than a vacuum (negative mass) could be created.

Rather than surpassing the speed of light within a local reference frame, a spacecraft using the Alcubierre Drive would traverse distances by contracting space in front of it and expanding space behind it, resulting in effective faster-than-light travel. Objects cannot accelerate to light speed within normal spacetime. Instead, the Alcubierre Drive alters the space around an object, allowing it to arrive at its destination faster than light would in regular space, without violating physical laws.

While the metric proposed by Alcubierre is consistent with Einstein's field equations, the practical construction of such a drive remains speculative. The mechanism requires negative energy density and, therefore, exotic matter or the manipulation of dark energy. If such exotic matter does not exist, building the drive is impossible. However, Alcubierre argued that the Casimir vacuum between parallel plates could satisfy the negative-energy requirement for the drive, as suggested by physicists studying traversable wormholes. Research indicates that the concept might be feasible with purely positive energy using 'soliton' waves.

In conclusion, subspace represents a fascinating and complex concept that bridges science fiction and theoretical physics. It challenges our understanding of space, time, and the fundamental forces of the universe, offering a tantalizing glimpse into the possibilities of faster-than-light travel and communication.

Chapter 62.1: In 1994, Miguel Alcubierre

In 1994, theoretical physicist Miguel Alcubierre proposed a revolutionary concept: altering the geometry of space itself to enable faster-than-light travel. He theorized that by creating a wave, the fabric of space ahead of a spacecraft could be contracted, while the area behind it expanded. This would form a region of flat space, known as a warp bubble, in which the spacecraft could ride. Within this warp bubble, the spacecraft would not move in the conventional sense but would be transported as the region itself moved due to the actions of the driver.

Mathematics of the Warp Drive

Using the Arnowitt-Deser-Misner (ADM) formalism of general relativity, spacetime is described by a foliation of space-like hypersurfaces of constant coordinate time tt, with the metric taking the following form:

$$Ds^2 = -(\alpha^2 - \beta_i\beta^i)dt^2 + 2\beta_i dx^i dt + \gamma_{ij}dx^i dx^j$$

Where:

- A is the lapse function, providing the interval of proper time between nearby hypersurfaces.
- β_i is the shift vector, relating spatial coordinate systems on different hypersurfaces.
- γ_{ij} is a positive-definite metric on each of the hypersurfaces.

The specific form that Alcubierre studied is defined by:

- $\alpha = 1$
- $\beta^x = -v_s(t)f(r_s(t))$
- $\beta^y = \beta^z = 0$
- $\gamma_{ij} = \delta_{ij}$

Where:

- $v_s(t) = \dfrac{dx_s(t)}{dt}$
- $r_s(t) = \sqrt{(x - x_s(t))^2 + y^2 + z^2}$
- $f(r_s) = \dfrac{\tanh(\sigma(r_s+R))-\tanh(\sigma(r_s-R))}{2\tanh(\sigma R)}$

With arbitrary parameters R>0 and σ > 0, Alcubierre's metric can be expressed as:

$$Ds^2 = \left(v_s(t)^2 f(r_s(t))^2 - 1\right)dt^2 - 2v_s(t)f(r_s(t))dxdt + dx^2 + dy^2 + dz^2$$

This particular form of the metric shows that the energy density measured by observers whose 4-velocity is normal to the hypersurfaces is given by:

$$-\frac{c^4}{8\pi G}\frac{v_s^2 y^2 + z^2}{4gr_s^2}\left(\frac{df}{dr_s}\right)^2$$

Where g is the determinant of the metric tensor. Due to the negative energy density, exotic matter is required to achieve faster-than-light travel.

Practical Considerations

The existence of exotic matter is theoretically not ruled out, but generating and sustaining sufficient exotic matter to perform feats such as faster-than-light travel and maintaining the "throat" of a wormhole is thought to be impractical. Writer Robert Low has argued that constructing a warp drive within the framework of general relativity is impossible without exotic matter.

Conclusion

Alcubierre's proposal, while mathematically consistent with Einstein's field equations, remains a speculative idea. The challenge lies in finding or creating exotic matter with the necessary properties. If achieved, the Alcubierre Drive would represent a monumental leap in space travel, potentially allowing humanity to traverse the stars at unprecedented speeds.

Chapter 62.2: Physics

The Alcubierre metric introduces unique peculiarities concerning specific effects of special relativity, such as Lorentz contraction and time dilation. One of the most intriguing aspects of the Alcubierre drive is that a ship using this drive would travel along a free-fall geodesic, even while the warp bubble is accelerating. This means that the crew on board would experience free fall during acceleration, avoiding the usual accelerational g-forces that accompany rapid movement.

Special Relativity Effects

Lorentz contraction and time dilation are fundamental effects of special relativity. Lorentz contraction refers to the shortening of objects in the direction of their motion as they approach the speed of light, while time dilation describes the phenomenon where time appears to slow down for objects moving at relativistic speeds. The Alcubierre drive, however, operates in a manner that allows the ship to avoid these relativistic effects by warping the space around it.

Free-Fall Geodesic

In general relativity, a geodesic is the shortest path between two points in curved spacetime, analogous to a straight line in flat space. The Alcubierre drive enables a spacecraft to travel on a free-fall geodesic, meaning the ship moves without the influence of external forces. This remarkable characteristic ensures that the crew remains in a state of free fall, even during acceleration, which would typically produce significant g-forces.

Tidal Forces

Despite the advantageous free-fall conditions within the warp bubble, the Alcubierre metric is not without its challenges. Enormous tidal forces would be present near the edges of the flat-space volume due to the intense curvature of space. These tidal forces arise from the gradient in gravitational strength across different points in space, which can stretch and compress objects.

To mitigate these effects, a carefully designed metric specification is essential. By fine-tuning the parameters of the warp bubble, it is possible to minimize tidal forces within the volume occupied by the spacecraft, ensuring the safety and comfort of the crew.

Conclusion

The physics of the Alcubierre drive presents a fascinating blend of general relativity and innovative theoretical concepts. While the notion of faster-than-light travel remains speculative, the insights gained from studying the Alcubierre metric offer intriguing possibilities for future advancements in space travel. By addressing challenges such as tidal forces and optimizing metric specifications, we inch closer to realizing the dream of traversing the cosmos at unprecedented speeds.

Chapter 63.0: The City on the Edge of Forever (Episode)

"The City on the Edge of Forever" stands as the twenty-eighth and penultimate episode of the first season of the American science fiction television series *Star Trek*. Harlan Ellison wrote the original teleplay, but contributors and editors to the script included Steven W. Carabatsos, D. C. Fontana, and Gene L. Coon. Gene Roddenberry made the final rewrite. The episode was directed by Joseph Pevney and first aired on NBC on April 6, 1967.

Plot Summary

In this episode, Doctor Leonard McCoy (DeForest Kelley), under the influence of a powerful medication, accidentally travels back in time and alters history. Captain James T. Kirk (William Shatner) and First Officer Spock (Leonard Nimoy) follow him through a time portal to rectify the timeline. As they navigate the past, Kirk falls deeply in love with Edith Keeler (Joan Collins), a compassionate and idealistic social worker. However, he soon realizes that to restore the future, he must allow her to die.

Discovery of the Portal

The narrative begins with the USS Enterprise encountering disturbances in the fabric of time. Kirk, Spock, Scott, Uhura, Galloway, and a security officer transport down to a ruined city, where they discover an enigmatic portal. While Uhura and Scott's teams search for Dr. McCoy, Kirk and Spock investigate the portal. Spock detects that it is the epicenter of the time disruptions, though the nature of its operation remains a mystery. The seemingly inert object emits powerful waves of displacement detectable by the Enterprise millions of miles away.

The Guardian of Forever

Curious about the object, Kirk asks, "What is it?" A loud, booming voice responds, "A question! Since before your sun burned hot in space and before your race was born, I have awaited a question." The entity introduces itself as the Guardian of Forever, explaining that it is both a machine and a being, yet also neither. Spock concludes that the Guardian is a time portal—a gateway to other times and dimensions. The Guardian corroborates this and activates its portal, presenting Kirk and Spock with an opportunity to journey into Earth's past.

Temporal Adventure

Venturing through the portal, Kirk and Spock arrive in 1930s New York City, a time of economic depression and societal struggle. They quickly locate McCoy but discover that his presence has altered the course of history. During their mission, Kirk meets Edith Keeler, a visionary who dreams of a better future. Despite their growing affection, Spock's analysis reveals that Edith's survival leads to a timeline where Nazi Germany wins World War II, resulting in a dystopian future.

The Heart-Wrenching Decision

Kirk faces an agonizing dilemma: to save the future, Edith must die. As events unfold, McCoy unwittingly prevents Edith from being hit by a car. Kirk, understanding the dire consequences, intervenes, allowing the tragic accident to occur. The timeline is restored, but Kirk is left heartbroken by the loss of Edith, who represented his idealistic dreams and aspirations.

Legacy and Impact

"The City on the Edge of Forever" is often hailed as one of the greatest episodes of *Star Trek*, praised for its compelling narrative, emotional depth, and philosophical themes. It explores the complexities of time travel, the moral implications of altering history, and the poignant sacrifices required for the greater good. The episode remains a timeless classic, resonating with audiences and inspiring future generations of storytellers.

Chapter 64.0: Different Types of Philosophy

The philosophy of space and time explores the fundamental issues surrounding the ontology, epistemology, and character of space and time. These concepts have been central to philosophical inquiry since its inception and have greatly influenced early analytic philosophy.

Epistemology

Epistemology, the theory of knowledge, examines the methods, validity, and scope of knowledge. It distinguishes justified belief from mere opinion. Epistemology seeks to answer questions such as: How do we know what we know? What constitutes true knowledge? What are the limits of human understanding?

Ontology

Ontology is a branch of metaphysics concerned with the nature of being. It involves a set of concepts and categories that demonstrate the properties and relations within a subject area or domain. Ontology seeks to understand what exists, the nature of existence, and how entities within a given domain relate to one another.

The Gap Between Knowledge and Wisdom

This philosophical inquiry examines whether the scientific conception of the world negates the need to search for the meaning of life. Ancient philosophy often intertwined knowledge and wisdom, suggesting that understanding the nature of things would lead to a better understanding of how to live one's life. However, the extraordinary progress in the sciences has created a separation between these two aspects. The question of wisdom, and the related question of the meaning of life, should ideally become more central to philosophical activity.

Scientism vs. Obscurantism

The debate between scientism and obscurantism addresses the role of science and technology in human alienation from the world. Scientism holds that the theoretical or natural-scientific way of viewing things provides the primary and most significant understanding of our world. However, this perspective can overlook the importance of human experiences and the life-world context. Anti-scientism does not reject science but argues that the practices of natural sciences arise from life-world methods and are not wholly reducible to scientific explanation.

Philosophy of Space and Time

This branch of philosophy delves into the ontological and epistemological questions surrounding space and time. How do we conceptualize these dimensions? Are space and time fundamental aspects of reality, or are they constructs of human perception? This field has inspired numerous philosophical debates and continues to be a vital area of inquiry.

Conclusion

Philosophy encompasses a wide range of inquiries into the nature of knowledge, existence, and reality. By exploring different branches, such as epistemology and ontology, and engaging in debates like scientism versus obscurantism, philosophy seeks to bridge the gap between knowledge and wisdom. Ultimately, it aims to provide a deeper understanding of the world and our place within it.

Chapter 64.1: Analytic Philosophy and Continental Philosophy

Analytic Philosophy

Analytic philosophy approaches philosophical problems by meticulously analyzing terms, a practice rooted in early 20th-century Anglo-American philosophy. It emphasizes language, a shift known as the linguistic turn, and is celebrated for its clarity and rigor in arguments, utilizing formal philosophical analysis.

Analytic philosophy is founded on the idea that philosophical problems can be resolved through precise terminology and systematic logic. This approach spans all significant branches of philosophy, including social and political philosophy, metaphysics, and logic. This movement, which emerged in the 20th century, advocates that philosophy should employ analytical techniques to achieve conceptual clarity, aligning itself with the successes of modern science.

Continental Philosophy

Continental philosophy represents a distinct set of traditions and practices that often address issues overlooked by the analytic tradition. Philosophers associated with this tradition include Immanuel Kant, Georg Wilhelm Friedrich Hegel, Friedrich Nietzsche, Edmund Husserl, Martin Heidegger, Jean-Paul Sartre, Jürgen Habermas, Michel Foucault, and Jacques Derrida. Central to Continental philosophy are critical concepts such as existentialism, nihilism, and phenomenology, which are thoroughly explained within this tradition.

Continental philosophy explores a broad range of topics, from existential and phenomenological inquiries to post-structuralism and critical theory. It delves into the human condition, the structures of consciousness, and the dynamics of power and society. By addressing these complex issues, Continental philosophy provides a rich and diverse philosophical landscape.

The Divide and Reconciliation

The divide between Analytic and Continental philosophy stems from differing methodologies, priorities, and philosophical goals. Analytic philosophy prioritizes logical clarity and formal analysis, while Continental philosophy embraces a more interpretive and critical approach. Despite these differences, there is a growing recognition of the need to bridge this divide, fostering a more integrated and comprehensive philosophical dialogue.

Continental Philosophy: A Very Short Introduction highlights the reasons for the conflict between these traditions and argues for overcoming this divide. It suggests that philosophers should embrace the strengths of both traditions to address a wider array of philosophical problems.

Conclusion

Analytic and Continental philosophies offer distinct yet complementary approaches to philosophical inquiry. Analytic philosophy's emphasis on language and logic provides clarity and precision, while Continental philosophy's exploration of existential, phenomenological, and critical issues offers depth and breadth. By recognizing and integrating the strengths of both traditions, philosophers can achieve a richer and more nuanced understanding of the complex questions that shape our world.

Chapter 64.2: Existentialism, Nihilism, and Phenomenology

Existentialism

Existentialism is a philosophical inquiry that delves into the complexities of human existence. It focuses on the experiences of thinking, feeling, and acting. Existentialist philosophers explore the nature of individual freedom, the search for meaning, and the inherent challenges of human life. Key figures in existentialism include Jean-Paul Sartre, who emphasized the concept of existential angst and the responsibility that comes with freedom, and Søren Kierkegaard, who explored the individual's relationship with God and the importance of personal choice.

Existentialism often grapples with the idea that life may lack inherent meaning, prompting individuals to create their own purpose through authentic choices and actions. The philosophy encourages people to confront the absurdity of existence and to live genuinely despite the absence of predetermined meaning.

Nihilism

Nihilism is a philosophical stance that rejects general or fundamental aspects of human existence, such as objective truth, knowledge, morality, values, or meaning. Nihilistic thought posits that life is inherently meaningless and that traditional beliefs and values are baseless. Friedrich Nietzsche is a prominent figure associated with nihilism, particularly his assertion that the "death of God" leads to a profound crisis of values.

Nihilism can take various forms, including existential nihilism, which focuses on the futility of human existence; moral nihilism, which denies the existence of absolute moral values; and epistemological nihilism, which doubts the possibility of knowledge. Despite its seemingly bleak outlook, nihilism can also serve as a catalyst for re-evaluating and reconstructing personal beliefs and values.

Phenomenology

Phenomenology is the philosophical study of the structures of experience and consciousness. Founded in the early 20th century by Edmund Husserl, phenomenology seeks to explore how individuals perceive and interpret the world around them. Husserl's followers later expanded the movement at the universities of Göttingen and Munich in Germany.

Phenomenologists aim to describe experiences as they are perceived, without preconceived notions or interpretations. This approach involves "bracketing" or suspending judgment about the external world to focus on the immediate experience. Martin Heidegger, a prominent student of Husserl, further developed phenomenology by examining the nature of being and the human experience of time and existence.

Phenomenology has influenced various fields, including psychology, sociology, and the cognitive sciences, by providing insights into the subjective aspects of human experience and consciousness.

Conclusion

Existentialism, nihilism, and phenomenology represent three distinct yet interconnected approaches to understanding the human condition. Existentialism explores the search for meaning and individual freedom, nihilism challenges the foundations of traditional beliefs and values, and phenomenology examines the structures of experience and consciousness. Together, these philosophical perspectives offer rich and diverse insights into the complexities of human existence.

Chapter 65.0: Why Will the Universe Remain a Mystery?

The great German philosopher Immanuel Kant offers profound insights into why the universe may always remain a mystery to us. According to Kant, our perception of reality is determined by the nature of our mind and our perceptive apparatus. By analogy, a computer's instruction set limits its computational capabilities. Similarly, our mental instruction sets constrain our ability to comprehend the full scope of reality.

Kant's brilliance lies in his understanding that the limits of our perception shape our understanding of the universe. I've often found myself quoting Kant, recognizing the profundity of his ideas but struggling to fully grasp their meaning. Today, I feel a sense of clarity regarding his assertion that our mental instruction sets bound our comprehension. These instructions, in essence, define the limits of our cognitive and perceptual capabilities.

The Source of Our Instructions

The question arises: How do we acquire these instructions? Who imparts them to us? Some might argue that these instructions are divine, delivered by God through sacred texts like the Bible. For the sake of argument, let's assume this theistic perspective. Even then, how are these instructions transmitted to us, and how do they encode survival knowledge into a species?

Throughout history, there are no records of God providing comprehensive biological instructions. For instance, the guidance given to Moses focused on social, rather than biological, instructions. Yet, animals innately know when to hunt, what food is edible, and how to navigate their environment. These instincts and behaviors must stem from some source, beyond mere magic.

Kant and Darwin: Pioneers of Understanding

Immanuel Kant and Charles Darwin are two fundamental thinkers who laid the groundwork for our understanding of the universe. Darwin's theory of evolution establishes that things change over time, highlighting the dynamic nature of life. On the other hand, Kant's philosophy emphasizes the limits of our perception and comprehension.

Our knowledge is deeply rooted in our environment, shaped by the Earth and its ever-changing conditions. Our sensory systems are designed to receive and respond to stimuli from the Earth, enabling us to navigate and survive in our surroundings. Animals, for example, use natural navigation systems to migrate during specific seasons in search of food, water, or mating opportunities. They rely on an inner compass, mental maps, and other cues to guide their journeys.

The Great Migration of Africa

One of the most remarkable examples of animal migration is the Great Migration of Africa. Thousands of wildebeests, zebras, and other species embark on an arduous journey across the African plains, driven by the instinctual knowledge encoded within them. These migrations are guided by both internal and external stimuli, demonstrating the profound interplay between an organism's instruction set and its environment.

Conclusion

Ultimately, the universe may remain a mystery because our understanding is limited by our perceptual and cognitive constraints. Kant and Darwin's insights provide a foundation for recognizing these limitations and appreciating the complexity of life and the cosmos. As we continue to explore and question, we inch closer to unraveling the mysteries of existence, even if we never fully comprehend them.

Chapter 65.1: The Rut

By June, the rains cease, and the dry season in Tanzania propels the herds further north. During this period, smaller groups amalgamate into larger flocks, and the migrating wildebeest, known for their rutting behavior, enter their mating season. Over a span of four weeks, upwards of 500,000 cows will mate. August and September offer the best opportunities to witness the great migration, with heightened chances of observing a river crossing. Come early October, the rainy season in Tanzania begins, prompting the herds to embark on their return journey to the Serengeti.

Adaptation and Sensory Systems

These migratory instructions are a direct response to the changing conditions of the Earth. It is fair to assert that animals have developed intricate systems to recognize and react to these environmental stimuli for survival and navigation. Some animals possess sensory systems so finely attuned to their surroundings that their very survival hinges on this acute perception.

Kant's brilliance becomes evident through observation. We can deduce that if any sensory input were too detached, it would limit the organism's other senses. What if a species were born without one of these senses? Their perception of the world would change dramatically. Evolution might eliminate certain senses, altering a species' reality of the universe. Consequently, scientists' calculations and models of the universe might never be entirely accurate, as the new species would lack the senses to detect what was once perceivable.

The Complexity of Perception

Scientists often define 'nothing' as having eternal or timeless features—simple, empty, plain, quiet, or even perfect symmetry, implying non-existence. Alternatively, 'nothing' might be considered beyond all existential description, in which case it was never simple and never will be.

Conclusion

Our understanding of the universe is profoundly limited by our sensory and cognitive capacities. As environmental conditions change, species adapt their sensory systems to survive and navigate their surroundings. Kant and Darwin's insights provide a foundation for recognizing these limitations and appreciating the complexity of life and the cosmos. Despite our best efforts, the universe's true nature may always remain a mystery, shaped by the ever-evolving interplay between perception and reality.

Chapter 65.2: The First Premise Scientists Make

The first premise scientists often make is that everything in the universe exists on Earth. Since we perceive and understand the world through our senses, and since our sensory systems have evolved to detect Earth's stimuli, our definitions and understandings are inherently limited. When we investigate space and encounter apparent nothingness, we are temporarily debarred from perceiving beyond our earthly sensory input. Not everything in the universe exists on Earth, and our senses are designed specifically to respond to terrestrial stimuli.

The Matrix Analogy

Consider the concept of the Matrix. In this scenario, machines have taken over the world and subjected humans to a program called the Matrix. This program intercepts signals from human senses, interprets them, and replaces them with deceptive signals, thereby controlling human reality and determining their existence.

Over time, humans evolve to adapt to this new reality. They develop specialized organs and discard those that are no longer necessary. These newly formed organisms still resemble humans but operate differently, as their sensory systems have evolved to accept only signals from the Matrix, not the natural environment. When disconnected from the Matrix using the red pill, a human would perceive nothingness and recognize only a few shared sensory inputs. Upon gaining consciousness, their evolved senses would detect only the transferred signals, discarding environmental signals. This limited perception would shape their reality, confined to the observable universe they could detect.

The Universe Beyond Our Senses

In reality, the universe exists in limitless form, containing structures and phenomena beyond our sensory capabilities. Consider an advanced micro-civilization that evolves to live inside a human host. This species undergoes a unique evolutionary process to survive within the human body. The host's environmental program influences the micro-host's instructional set, designed for internal survival.

Star Trek Voyager Analogy

Using a similar analogy from the famous *Star Trek: Voyager* episode "Blink of an Eye" (the twelfth episode of the sixth season), we see the crew interacting with a world where time passes rapidly. The USS Voyager remains fixed in the night sky, inspiring the inhabitants as eons pass. This concept highlights the real-world physical effect of time dilation, which must be considered in satellite communication signals from Earth's orbit.

Perception and Reality

Eventually, the micro-species inside the host body reaches the body's surface and looks out into the vast expanse of space. They perceive nothing because their senses are not programmed to detect environmental signals, as it is not crucial for their survival. This limited perception shapes their understanding of reality.

Their sensory systems evolved to meet the needs of their environment, which means they can only detect heat and light. Gazing into the universe, they identify these signals but nothing else. These specifications are tailored for life inside a human host. They could be hit by a moving car and remain unaware until it is too late. This illustrates that our sensory systems are designed to operate within Earth's parameters, not the universe's.

Conclusion

Our comprehension of the universe is profoundly influenced by our sensory and cognitive limitations. As species evolve, their sensory systems adapt to their environment, shaping their perception of reality. The universe's true nature may remain elusive, with our understanding confined by the interplay between perception and reality. Through the insights of philosophers like Kant and Darwin, we recognize these limitations and continue to explore the mysteries of existence.

Chapter 65.3: Our Perception of the Universe

When we gaze at the universe, we observe celestial bodies like the Earth, the Moon, and the Sun suspended in space, held in place by forces that maintain their balance. However, our understanding of what lies between these objects is limited. We perceive a vast expanse of dark black space, yet this might not be the true nature of the universe. Our sensory limitations and reasoning constrain what we can detect and comprehend.

The Limits of Our Senses

Our senses are specifically designed to respond to stimuli from Earth. As a result, we may be blind to other elements or energy signatures that exist in the infinite expanse of the universe. There could be entire planets or phenomena between Earth and Mars that we are unable to perceive because our sensory systems are not equipped to detect them.

The Matrix Analogy Revisited

Drawing from the Matrix analogy, imagine that our perception of reality is controlled by an external program. In this scenario, our senses are intercepted and manipulated, creating a deceptive reality. If humans were subjected to such a program, they would evolve to develop specialized sensory organs that only respond to the manipulated signals. Once disconnected from the Matrix, humans would perceive nothingness, as their evolved senses would disregard environmental signals.

This analogy highlights that our understanding of the universe is limited to the signals our senses can detect. Just as the Matrix controls human reality, our sensory systems control our perception of the universe.

The Star Trek Voyager Analogy

A similar concept is explored in the *Star Trek: Voyager* episode "Blink of an Eye," where the crew interacts with a rapidly evolving world. The inhabitants perceive Voyager as a stationary object in the sky, inspiring them as they progress through their history. This idea illustrates the real-world physical effect of time dilation, which must be considered when analyzing communication signals from satellites in Earth's orbit.

The Infinite and Unseen Universe

The universe exists in limitless form, containing elements and energy signatures beyond our sensory capabilities. An advanced micro-civilization evolving within a human host might only detect heat and light. Their sensory systems, tailored for survival inside the host, would not pick up signals from the external environment. This limited perception would shape their reality, much like our limited senses shape our understanding of the universe.

Conclusion

Our perception of the universe is profoundly influenced by the limitations of our senses and cognitive capacities. As species evolve, their sensory systems adapt to their environment, shaping their perception of reality. The true nature of the universe may remain elusive, with our understanding confined by the interplay between perception and reality. Recognizing these limitations allows us to continue exploring the mysteries of existence with an open mind.

Chapter 66.0: The Earth's Internet: How Fungi Help Plants Communicate

Fungi aren't just the internet systems of our world; they represent a profound and complex network through which plants communicate and share resources. This chapter delves into the microbial fungi methods of connection and how plants utilize this subterranean internet system, which is not only found on Earth but also has potential parallels in other network systems discovered by scientists.

The Invisible Network

The human internet connects more than half of the world's population through a vast and invisible web of servers, computers, and devices. It has revolutionized our lives, enabling interaction and access to immense information. However, humans are not the only organisms with an invisible interconnected network. Plants, though seemingly isolated and solitary, can communicate over considerable distances thanks to their unique relationship with fungi.

The Mycorrhizal Network

All plant species share a mutually beneficial relationship with soil fungi called mycorrhizae. These fungi grow a network of small, branching tubes called mycelium, which extend throughout the soil and interact with plant roots. This mycelial network allows fungi to absorb essential nutrients from the ground, such as nitrogen and phosphorus, which plants struggle to extract on their own. In exchange for these hard-to-obtain nutrients, plants trade carbon in the form of sugars with the fungi. This barter system enables both plants and fungi to thrive in environments where they might otherwise struggle.

Discovering the Underground Network

The symbiotic relationship between plants and fungi was first discovered in the early 1900s, but it wasn't until 1997 that the depth of this underground network became fully understood. Ecologist Suzanne Simard hypothesized that plants were not just sharing nutrients with fungi but also with each other. To test this, Simard and her colleagues infused trees in a forest with a traceable radioactive form of carbon. Later, they took samples from neighboring trees and found that many of them also contained radioactive carbon. This experiment proved that plants could send nutrients back and forth, distributing them where they were needed most.

Communication Through the Mycorrhizal Network

This discovery revealed that plants use the mycorrhizal network to communicate and share resources, much like how the internet facilitates communication and information sharing among humans. Through this network, plants can warn each other of pests, share nutrients, and support the growth of seedlings. The mycorrhizal network essentially acts as an underground internet, allowing plants to interact in ways that were once thought impossible.

Similar Network Systems Beyond Earth

Interestingly, scientists have discovered similar network systems beyond Earth. These findings suggest that the principles of interconnectedness and resource sharing may be fundamental to life itself, extending beyond our planet. The study of these networks can provide valuable insights into the nature of life and the potential for communication and cooperation in different environments.

Conclusion

The Earth's internet, formed by the intricate mycorrhizal network, demonstrates the incredible complexity and interconnectedness of life. Plants and fungi engage in a symbiotic relationship that enables them to communicate and thrive in their environments. This underground network parallels the human internet in its ability to facilitate interaction and resource sharing. By understanding and appreciating these natural systems, we gain a deeper insight into the fundamental connections that sustain life on Earth and beyond.

Chapter 66.1: The Mycelial Network and Plant Communication

Plants rely on light energy to convert carbon dioxide and water into sugar and oxygen through the magical process of photosynthesis. However, plants in the shade receive less light and, consequently, produce less sugar. Ecologist Suzanne Simard discovered that these shaded, energy-deficient trees had more radioactive carbon than their sunlit counterparts, indicating that the plant-fungi relationship is akin to feeding the hungry.

Mycelial Networks: Nature's Communication System

Research into mycelial networks has revealed that plants gain more nutrients and engage in sophisticated communication through these underground systems. Plants "talk" chemically through mycelia, sharing essential information and resources. Seedlings connected to the Common Mycelial Network (CMN) have a higher chance of survival, and plants that are "online" are usually healthier. This health boost is attributed to an early warning system facilitated by the CMN.

When a plant is under attack, it releases chemicals that alert nearby plants to impending danger. This communication occurs through both airborne compounds and the CMN. Other plants heed this warning and activate their defenses before the threat reaches them. For instance, when a pest attacks tomato plants connected by a CMN, neighboring plants will bolster their defenses in response.

The Interconnected Forest

Scientists are only beginning to comprehend the critical role these plant networks play. They have discovered that entire forests can be interconnected, though connectivity varies unevenly, much like the human internet. Older, larger trees, known as hubs or mother trees, are more connected than others. These trees possess extensive root networks that host a greater diversity of mycorrhizal fungi, enabling them to interact with many other plants.

Mother trees play a crucial role in forest ecosystems. They can send "care packages" of extra nutrients to their kin, helping them survive and thrive. This nurturing behavior earns them the moniker "mommy trees." During times of change, mother trees help forests transition. When they are injured or dying, they release a surge of carbon into the network, nurturing the next generation of trees, even if they belong to different species.

The Dark Side of Plant Communication

No network is without its hackers. Some plants claim territory and influence community dynamics by releasing toxins into the CMN. For example, black walnuts release toxins into the soil through these networks. Plants immune to the toxins thrive, while others struggle or die off. Harmful worms, parasitic plants, and fungi can exploit the mycelial network to find their target plants by following the underground chemical trails.

Conclusion

The mycelial network demonstrates the intricate and interconnected nature of plant communication and resource sharing. Through this underground network, plants can warn each other of threats, share nutrients, and support the growth of new generations. The discovery of these networks offers profound insights into the complexity of life on

Earth and the sophisticated ways in which organisms interact and thrive. Understanding these natural systems can inspire new approaches to sustainability and ecological balance.

Chapter 66.2: Connecting with the Earth's Internet

It's astounding to consider that this chemical information superhighway has existed right beneath our noses for eons, yet we remained oblivious. Now, with a better understanding, we can finally plug into this network and use it to connect more constructively with the planet's flora. Knowledge of this interconnectivity is revolutionizing our relationship with plants, aiding forest conservation and agriculture.

The Role of Mycorrhizal Networks in Conservation and Agriculture

Preserving highly connected mother trees during deforestation efforts ensures the diversity of mycorrhizal fungi and promotes rapid forest regrowth. These mother trees are crucial hubs in the mycelial network, facilitating nutrient distribution and enhancing ecosystem resilience. In agriculture, farming in soil rich with a Common Mycelial Network (CMN) allows plants to communicate and warn each other of invading pests. This natural defense mechanism can reduce the need for chemical pesticides, promoting a healthier and more sustainable environment.

The Unseen Power of Fungi

Fungi are remarkable organisms. A brief exploration into their world reveals their essential role in decomposition and nutrient recycling within ecosystems. Fungi are indispensable to every ecosystem, existing worldwide and often remaining unstudied unless they directly impact human interests. They are a critical component of biodiversity, influencing ecological balance in ways that most people are unaware of.

Enhancing Security and Knowledge

Just like the human internet, the Earth's mycelial network increases security and knowledge for its connected members. Plants that are part of this network can share information about threats, nutrient availability, and environmental changes. This communication enhances the survival and health of individual plants and entire ecosystems.

Conclusion

Recognizing and utilizing the Earth's internet can transform how we interact with nature. By preserving key components of this network, such as mother trees, and promoting mycorrhizal connectivity in agriculture, we can foster more sustainable and resilient ecosystems. Fungi, as a vital part of this network, play an indispensable role in maintaining ecological balance and supporting plant life. Embracing this knowledge can lead to more informed and effective conservation and agricultural practices, benefiting both the environment and human well-being.

Chapter 67.0: Fungi Are Responsible for Life on Land as We Know It!

Neither plants nor animals, fungi are perhaps the most underappreciated kingdom of the natural world. Over a billion years of evolution, fungi have become masters of survival and have played an integral role in the development of life on Earth. Without them, neither land plants nor terrestrial animals would exist.

The Unique Kingdom of Fungi

Mold, a living organism belonging to the kingdom of Fungi, exemplifies the unique nature of fungi. Although some fungi appear plant-like, they are neither plant nor animal. Mold is heterotrophic, meaning it cannot produce its own food like plants. Instead, it must obtain nutrients from other organic substances.

Symbiosis with Plants

Fungi have several characteristics in common with plants: they have cell walls, vacuoles, and reproduce by both sexual and asexual means. Like basal plant groups such as ferns and mosses, fungi were among the first complex life forms on land. They mined rocks for mineral nourishment, gradually transforming them into soil. During the Late Ordovician era, fungi formed a symbiotic relationship with liverworts, the earliest plants.

"Ultimately, fungi helped plants move from these marginal tiny things on the water's edge into large forests and entire ecosystems," explains Katie Field, an associate professor in plant-soil interactions. The fungi provided essential minerals that enabled land plants to spread and turn the planet green, altering the composition of the atmosphere.

Fungi and Catastrophic Events

Approximately 65 million years ago, an asteroid strike wiped out 70 percent of all life on Earth. However, this extinction event did not occur instantly. The ensuing lack of sunlight caused plant life to decay rapidly, creating ideal conditions for fungi to proliferate.

During this period, mammals had a significant advantage over cold-blooded reptiles, the dominant life forms at the time. "They're hot," explains Arturo Casadevall, a professor of public health at Johns Hopkins University. "Reptiles are quite susceptible to fungal diseases, but typical mammals, which maintain a temperature in the mid-'30s, create a thermal exclusionary zone for fungi."

The surviving mammals are the evolutionary ancestors of every mammal on the planet today, from civet cats to water buffaloes to humans. "The warm-bloodedness of mammals, including ourselves, has evolved, in part, as a response to the pressure from fungus," says Rob Dunn, a professor at North Carolina State University. "And so, we seem to have cooked out the fungal pathogens."

The Unexplored World of Fungi

Scientists estimate that there are over 5,000,000 species of fungi on Earth, yet we have only discovered about one percent. Professor Rob Dunn suggests that as we delve deeper into the world of fungi, there is no telling what other breakthroughs might be possible.

Conclusion

Fungi have been essential to the development and sustenance of life on land. Their unique characteristics and symbiotic relationships with plants have enabled them to play a vital role in transforming the planet's ecosystems. Understanding and appreciating the significance of fungi can lead to new insights and discoveries, shedding light on the intricate web of life on Earth.

Chapter 68.0: Slime Mold Simulations Used to Map Dark Matter Holding the Universe Together

The single-cell organism, slime mold (*Physarum polycephalum*), builds complex filamentary networks in its quest for food, finding near-optimal pathways to connect different locations. Interestingly, gravity shapes the universe in a similar manner, creating a vast cobweb structure of filaments that tie galaxies and clusters along washed-out bridges spanning hundreds of millions of light-years. The resemblance between these two networks—one crafted by biological evolution and the other by the primordial force of gravity—is uncanny.

Slime Mold-Inspired Algorithms

Researchers designed a computer algorithm inspired by slime mold behavior and tested it against a simulation of the universe's growth of dark matter filaments. A computer algorithm is essentially a recipe that tells a computer exactly what steps to take to solve a problem.

Mapping the Cosmic Web

The researchers applied the slime mold algorithm to data containing the locations of 37,000 galaxies mapped by the Sloan Digital Sky Survey, at distances corresponding to 300 million light-years. The algorithm produced a three-dimensional map of the underlying cosmic web structure.

Scientists then analyzed ultraviolet light from 350 quasars—distant cosmic flashlights that are the brilliant, black-hole-powered cores of active galaxies cataloged in the Hubble Spectroscopic Legacy Archive. This archive holds data from NASA's Hubble Space Telescope's spectrographs. The light from these quasars shines across space and through the foreground cosmic web, imprinting the telltale absorption signature of otherwise undetected hydrogen gas. The team analyzed this gas at specific points along the filaments. These target locations, far from the galaxies, allowed the research team to link the gas to the universe's large-scale structure.

Unveiling the Cosmic Web

"It's fascinating that one of the simplest forms of life enables insight into the very largest-scale structures in the universe," said lead researcher Joseph Burchett of the University of California (UC), Santa Cruz. By using the slime mold simulation to find the location of cosmic web filaments—including those far from galaxies—the team could then use the Hubble Space Telescope's archival data to detect and determine the density of the cool gas on the very outskirts of these invisible filaments. Scientists have detected signatures of this gas for several decades, and this study has proven the theoretical expectation that this gas comprises the cosmic web.

Conclusion

The application of slime mold-inspired algorithms to map the cosmic web highlights the profound interconnectedness of life and the universe. This innovative approach demonstrates that even the simplest forms of life can provide valuable insights into the largest-scale structures in the cosmos. As we continue to explore and understand these connections, we gain a deeper appreciation for the intricate web that holds the universe together.

Chapter 69.0: The Darker Side of Fungi

While fungi have many beneficial roles, there is also a darker side to these organisms—one that most people associate with decay, rot, and destruction. Some fungi have surged out of control due to environmental imbalances, leading to devastating consequences for various species.

Climate Change and Fungal Infections

Climate change has pushed amphibians into climates where they are more susceptible to *Batrachochytrium dendrobatidis* (Bd), a fungus causing the deadly disease chytridiomycosis. This disease has led to drastic declines in amphibian populations across Latin America, Australia, and the western United States. Similarly, in the United States, white-nose syndrome, a fungal infection, has wiped out millions of bats, leaving insect populations unchecked. This syndrome was transported from Europe and thrives in the higher temperatures that are becoming more common due to the climate crisis.

Human Impact

In India, an infection called mucormycosis, caused by exposure to mucor mold, has affected individuals with compromised immune systems, particularly those recovering from COVID-19. This black fungus infection has added to the challenges faced by healthcare systems already strained by the pandemic.

Signs of Environmental Imbalance

The harmful effects of fungi often indicate a system out of balance. For instance, the spread of chytridiomycosis and white-nose syndrome reflects broader ecological disruptions. These fungal outbreaks can be seen as symptoms of larger environmental issues that need addressing.

Understanding Fungi's Dual Nature

Fungi, like any part of the natural world, play both healing and harmful roles in our lives and ecosystems. Understanding the darker side of fungi is crucial for developing strategies to mitigate their negative impacts while appreciating their beneficial contributions.

Conclusion

The darker side of fungi underscores the importance of maintaining ecological balance and understanding the complex interactions within ecosystems. As we continue to study fungi, we can develop better approaches to manage their harmful effects and harness their potential for beneficial uses. By deepening our knowledge of fungi, we can ensure a healthier, more sustainable relationship with these fascinating organisms.

Chapter 70.0: The "Fine Structure Constant"

The fine structure constant—one of the world's critical numbers that appears precisely tuned for life to exist—might not be constant after all! This possibility has profound implications for our understanding of the universe and the fundamental forces that shape it.

The Importance of Physical Constants

Some physicists have explored the notion that if the dimensionless fundamental physical constants had sufficiently different values, our universe would be so radically different that intelligent life would not have emerged. These constants include the rationalized Planck's constant h, the gravitational constant G, the speed of light in vacuum cc, the electric constant ε_0, the fine-structure constant α, and the elementary charge e.

The Anthropic Principle

The Strong Anthropic Principle states that the fundamental constants acquired their respective values in such a way that there was sufficient order and elemental diversity for life to form, ultimately evolving the necessary intelligence to observe these constants. This perspective suggests that our universe is fine-tuned for intelligent life. The Weak Anthropic Principle, on the other hand, posits that we observe these constants because we exist in a universe where these values allow for our privileged perspective.

Murphy's Law and Physical Constants

Murphy's Law, often humorously quoted, states that anything that can go wrong will go wrong. While not a physical constant, it highlights the importance of considering the unexpected in scientific inquiry.

Key Physical Constants

Here are some of the critical physical constants that govern our universe:

- Newtonian constant of gravitation (G): 6.6742×10^{-11} m^3/s^2/kg
- Speed of light (c): 299,792,458 m/s
- Permeability of vacuum (μ_0): $1.25663706143592 \times 10^{-7} H/m$
- Permittivity of vacuum (ε_0): $8.85418781762039 \times 10^{-12}$ F/m
- Fine structure constant (α): $7.297352568 \times 10^{-3}$
- Rydberg constant (R_∞): $10,973,731.568525 m^{-1}$
- Electron charge (e): $1.60217653 \times 10^{-19} C$
- Planck constant (h): $6.6260693 \times 10^{-34}$ J·s
- Electron mass (m_e): $9.1093826 \times 10^{-31} kg$
- Proton mass (m_p): $1.67262171 \times 10^{-27}$ kg
- Proton-electron mass ratio (m_p/m_e): 1836.15267261
- Boltzmann constant (k): $1.3806505 \times 10^{-23}$ J/K
- Stefan-Boltzmann constant (σ): 5.670400×10^{-8} W·m^{-2}·K^{-4}
- Wien displacement law constant (b): 2.8977685×10^{-3} m·K

Conclusion

The notion that the fine structure constant might not be constant adds an intriguing layer to our understanding of the universe. By exploring the values and implications of these fundamental constants, we gain deeper insights into the delicate balance that allows for the existence of life and the observable universe. As scientific inquiry progresses, we continue to uncover the mysteries that shape our reality.